CAMBRIDGE LIBRARY COLLECTION

Books of enduring scholarly value

Physical Sciences

From ancient times, humans have tried to understand the workings of the world around them. The roots of modern physical science go back to the very earliest mechanical devices such as levers and rollers, the mixing of paints and dyes, and the importance of the heavenly bodies in early religious observance and navigation. The physical sciences as we know them today began to emerge as independent academic subjects during the early modern period, in the work of Newton and other 'natural philosophers', and numerous sub-disciplines developed during the centuries that followed. This part of the Cambridge Library Collection is devoted to landmark publications in this area which will be of interest to historians of science concerned with individual scientists, particular discoveries, and advances in scientific method, or with the establishment and development of scientific institutions around the world.

Popular Lectures and Addresses

William Thomson, Baron Kelvin (1824–1907), was educated at Glasgow and Cambridge. While only in his twenties, he was awarded the University of Glasgow's chair in natural philosophy, which he was to hold for over fifty years. He is best known through the Kelvin, the unit of measurement of temperature named after him in consequence of his development of an absolute scale of temperature. These volumes collect together Kelvin's lectures for a wider audience. In a convivial but never condescending style, he outlines a range of scientific subjects to audiences of his fellow scientists. The range of topics covered reflects Kelvin's broad interests and his stature as one of the most eminent of Victorian scientists. Volume 2 is mainly concerned with geology and was actually published last, in 1894. It includes additional lectures given between 1866 and 1893 that were not included in the other two volumes.

Cambridge University Press has long been a pioneer in the reissuing of out-of-print titles from its own backlist, producing digital reprints of books that are still sought after by scholars and students but could not be reprinted economically using traditional technology. The Cambridge Library Collection extends this activity to a wider range of books which are still of importance to researchers and professionals, either for the source material they contain, or as landmarks in the history of their academic discipline.

Drawing from the world-renowned collections in the Cambridge University Library, and guided by the advice of experts in each subject area, Cambridge University Press is using state-of-the-art scanning machines in its own Printing House to capture the content of each book selected for inclusion. The files are processed to give a consistently clear, crisp image, and the books finished to the high quality standard for which the Press is recognised around the world. The latest print-on-demand technology ensures that the books will remain available indefinitely, and that orders for single or multiple copies can quickly be supplied.

The Cambridge Library Collection will bring back to life books of enduring scholarly value (including out-of-copyright works originally issued by other publishers) across a wide range of disciplines in the humanities and social sciences and in science and technology.

Popular Lectures and Addresses

VOLUME 2:
GEOLOGY AND GENERAL PHYSICS

LORD KELVIN

CAMBRIDGE UNIVERSITY PRESS

Cambridge, New York, Melbourne, Madrid, Cape Town,
Singapore, São Paolo, Delhi, Tokyo, Mexico City

Published in the United States of America by Cambridge University Press, New York

www.cambridge.org
Information on this title: www.cambridge.org/9781108029780

© in this compilation Cambridge University Press 2011

This edition first published 1894
This digitally printed version 2011

ISBN 978-1-108-02978-0 Paperback

POPULAR LECTURES

AND

ADDRESSES

VOL. II

NATURE SERIES

POPULAR LECTURES

AND

ADDRESSES

BY

SIR WILLIAM THOMSON (BARON KELVIN)

P.R.S., LL.D., D.C.L., &c.

PROFESSOR OF NATURAL PHILOSOPHY IN THE UNIVERSITY OF GLASGOW, AND
FELLOW OF ST. PETER'S COLLEGE, CAMBRIDGE

IN THREE VOLUMES

VOL. II

GEOLOGY AND GENERAL PHYSICS

WITH ILLUSTRATIONS

London

MACMILLAN AND CO.

AND NEW YORK

1894

RICHARD CLAY AND SONS, LIMITED,
LONDON AND BUNGAY.

PREFACE.

Of the reason why Volume III. appeared before Volume II. nothing need be said here, as it was stated in the Preface to Volume III.

The first half of the present volume contains the papers relating to geological subjects referred to in that statement. The remainder of the volume consists of lectures and addresses on various subjects of general physics, given between 1866 and the end of 1893, which have not been included in Vols. I. or III.

KELVIN.

UNIVERSITY, GLASGOW.
February 2, 1894.

CONTENTS.

POPULAR LECTURES

AND

ADDRESSES

Popular Lectures and Addresses.

PROTECTION OF VEGETATION FROM COLD.

[Paper read before the Royal Society of Edinburgh, April 4, 1864.]

THE effect of dew in protecting vegetation every clear still night of summer was long ago pointed out by Dr. Wells; the correctness and acuteness of whose views on this subject have been generally recognised. The hypothesis recently put forth by Dr. Tyndall, that absorption of radiant heat by aqueous vapour in the atmosphere is an effective defence against destructive degrees of cold, and the ready acceptance yielded to it by some of our highest authorities in the popular promulgation of the truths of science, seems to render it necessary

to recall attention to Dr. Wells's admirable work.
In the first place, when Dr. Tyndall announces, as
a result of his experiments on radiant heat, that
" It is perfectly certain that more than ten per
cent. of the terrestrial radiation from the soil of
England is stopped within 10 feet of the surface of
the soil," by the absorption it suffers from aqueous
vapour ; it must be remarked that this absorption
cannot go on at the same rate through any great
thickness of air.   For at the same rate half the
radiant heat would be absorbed in 70 feet ; $\frac{3}{4}$ in
140 feet ; $\frac{7}{8}$ in 210 feet, and so on, which is inconsistent with known facts ; as, for instance, the
influence of clouds on terrestrial radiation.   Hence
the quality of rays which passes through the
lowest 10 feet of air suffers less than ten per cent.
of absorption in the next 10 feet ; and it is quite
certain that after passing through several times 10
feet of air, the radiant heat must, by having been
deprived of the part of it specially liable to absorption by aqueous vapour, be in a condition in which
not one per cent. is absorbed from it in its passage
through 10 feet of clear air.   If true vapour of

water really does exercise any influence in check-
ing, by its absorption, the loss of heat by radiation
from the earth's surface, it is, even in the most
humid conditions of optically clear atmosphere,
insufficient to prevent heavy dews by radiation
into space of the latent heat of the vapour from
which they are condensed. The quantity of heat
thus radiated into space through the clear moist
air close to the ground is so great that if instead
of being taken from the vapour it were taken
from the blades of grass, or other finer parts of
plants, it would leave them destroyed by frost.

In point of fact heat *actually is radiated* away
into very high terrestrial atmosphere and distant
interstellar air or æther, from the upper and
finer parts of living plants, in so great amount
every clear night of summer, that destruction
by frost could not be delayed for many hours
after sunset without a compensating supply of
heat from some extraneous source. This source,
on windy nights, is the thermal capacity of
the air whirled about, up and down, and among
the stems and leaves of the plants. On still

nights it is the latent heat of the vapour condensed into dew. This vapour is taken chiefly from the air engaged among the stems and leaves, which, in the case at least of fine grass, is all *nearly* at the same temperature as the leaves ; the temperature of the surface of these being of course rigorously the same as that of the air in contact with them. Thus the temperature of the leaves can never go *below* the *dew-point* of the air touching them, and any cooling which they experience *after* dew begins to deposit upon them is only equal to the lowering of the *dew-point*, occasioned by the amount of drying experienced by the air in consequence of the condensation of vapour out of it.

Clouds, as remarked first by Prévost, being practically opaque, prevent the surface of the earth from *tending* by radiation to a lower temperature than their own, which, unless they are very high, is generally not much colder than the dew-point of the lower air, but is at all events in general sufficiently warm to prevent the finest blades of grass from acquiring any very sensible dew, or to allow the general temperature of grass and the air engaged among it, even on the stillest night, to sink as low as the dew-point. Thus either clouds, by their counter radiation, or wind, by mixing a comparatively thick stratum of air with that next the earth, keep the grass and delicate parts of other plants from sinking to the dew-point. When there is not enough of clouds and wind to afford this degree of protection, dew begins to form, and by preventing the temperature of any leaf or flower from sinking below the dew-point, saves them all from destruction, unless, as when hoar-frost appears, the dew-point itself is below the freezing-point.

[Added *December* 15, 1892.]—Thus when neither clouds, nor wind blowing among the plants, suffice to protect them from sinking to the dew-point, the temperature to which leaves, flowers and grass sink is lower the dryer is the air, not because dry clear air is more diathermanous than moist clear air,[1] but because the dew-point is lower the dryer is the air.

[1] Tyndall, "On Terrestrial Radiation," *Proc. Roy. Soc.*, February, 1883.

THE
"DOCTRINE OF UNIFORMITY"
IN GEOLOGY BRIEFLY
REFUTED.

[*Paper read before the Royal Society of Edinburgh,*
December 18, 1865.]

THE "Doctrine of Uniformity" in Geology, as
held by many of the most eminent of British
Geologists, assumes that the earth's surface and
upper crust have been nearly as they are at present
in temperature, and other physical qualities, during millions of millions of years. But the heat
which we know, by observation, to be now conducted out of the earth yearly is so great, that if
this action had been going on with any approach
to uniformity for 20,000 million years, the amount
of heat lost out of the earth would have been about
as much as would heat, by 100° Cent., a quantity
of ordinary surface rock of 100 times the earth's
bulk. (See calculation appended.) This would be

more than enough to melt a mass of surface rock equal in bulk to the *whole earth.* No hypothesis as to chemical action, internal fluidity, effects of pressure at great depth, or possible character of substances in the interior of the earth, possessing the smallest vestige of probability, can justify the supposition that the earth's upper crust has remained nearly as it is, while from the whole, or from any part, of the earth, so great a quantity of heat has been lost.

APPENDIX.

ESTIMATE OF PRESENT ANNUAL LOSS OF HEAT FROM THE EARTH.

LET A be the area of the earth's surface, D the increase of depth in any locality for which the temperature increases by 1° Cent., and k the conductivity per annum of the strata in the same locality. The heat conducted out per annum per square foot of surface in that locality is $\dfrac{k}{D}$. Hence, if we give k and D proper average values for the whole upper crust of the earth, the quantity conducted out across the whole earth's surface per annum will

be $\dfrac{kA}{D}$. The bulk of a sphere being its surface

multiplied by $\frac{1}{3}$ of its radius, the thermal capacity
of a mass of rock equal in bulk to the earth, and of
specific heat s per unit of bulk is $\frac{1}{3}$ Ars.   Hence

$\dfrac{3k}{Drs}$ is the elevation of temperature which a

quantity of heat equal to that lost from the earth
in a year, would produce in a mass of rock equal
in bulk to the whole earth.   The laboratory experiments of Peclet; Observations on Underground Temperature in three kinds of rock in and
near Edinburgh, by Forbes ; in two Swedish strata,
by Ångström, and at the Royal Observatory,
Greenwich, give values of the conductivity in
gramme-water units of heat per square centimetre, per 1° per centimetre of variation of temperature, per second, from ·002 (marble, Peclet) to
·0107 (sandstone of Craigleith quarry, Forbes) ; and
·005 may be taken as a rough average.   Hence, as
there are 31,557,000 seconds in a year, we have
$k = ·005 \times 31,557,000$, or approximately 16×10^4.
The thermal capacity of surface rock is somewhere
about half that of equal bulk of water ; so that
we may take $s = ·5$.   And the increase of temperature downwards may be taken as roughly averaging 1° Cent. per 30 metres ; so, that, D = 3000
centimetres.   Lastly, the earth's quadrant being
according to the first foundation of the French

metrical system, about 10^9 centimetres, we may take, in a rough estimate such as the present, $r=6 \times 10^8$ centimetres. Hence,

$$\frac{3k}{Drs} = \frac{3 \times 16 \times 10^4}{3000 \times 6 \times 10^8 \times \cdot 5} = \frac{8}{15 \times 10^6}$$

This, multiplied by 20,000 × 10^6, amounts to 10,000, or to 100 times as much heat as would warm 100 times the earth's bulk of surface rock by 1° Cent.

ON GEOLOGICAL TIME.

[Address delivered before the Geological Society of Glasgow, February 27, 1868.]

1.—A GREAT reform in geological speculation seems now to have become necessary. A very earnest effort was made by geologists, at the end of last century, to bring geology within the region of physical science, to emancipate it from the dictation of authority and from dogmatic hypotheses. The necessity for *more time* to account for geological phenomena than was then generally supposed to be necessary, became apparent to all who studied with candour and with accuracy the phenomena presented by the surface of the earth. About the end of last century, also, physical astronomers made great steps in the theory of the motions of the heavenly bodies, and, among other remarkable propositions, the very celebrated theorem of the stability of the planetary motions was announced. That theorem was taken up somewhat rashly, and supposed to imply more than it really did with reference to the permanence of the solar system. It was probably it which Playfair had in his mind when he wrote that celebrated and often-quoted passage—" How often " these vicissitudes of decay and renovation have " been repeated is not for us to determine ; they " constitute a series of which, as the author of " this theory has remarked, we neither see the " beginning nor the end ; a circumstance that " accords well with what is known concerning other " parts of the economy of the world. In the " continuation of the different species of animals " and vegetables that inhabit the earth, we discern " neither a beginning nor an end ; in the planetary " motions where geometry has carried the eye so " far both into the future and the past, we discover "no mark either of the commencement or the " termination of the present order. It is unreason- " able, indeed, to suppose that such marks should " anywhere exist. The Author of nature has not

"given laws to the universe, which, like the
"institutions of men, carry in themselves the
"elements of their own destruction. He has not
"permitted in His works any symptoms of infancy,
"or of old age, or any sign by which we may
"estimate either their future or their past duration.
"He may put an end, as He, no doubt, gave a
"beginning to the present system, at some
"determinate time ; but we may safely conclude
"that this great *catastrophe* will not be brought
"about by any of the laws now existing, and
"that it is not indicated by anything which
"we perceive." (*Illustrations of the Huttonian
Theory*, § 118.) Nothing could possibly be
further from the truth than that statement. It is
pervaded by a confusion between "present order,"
or "present system," and "laws now existing"—
between destruction of the earth as a place
habitable to beings such as now live on it, and a
decline or failure of law and order in the universe.
The theorem of the French mathematicians
regarding the motions of the heavenly bodies is a
theorem of approximate application, and one

which professedly neglects frictional resistance of every kind ; and the statement that the phenomena presented by the earth's crust contain no evidence of a beginning, and no indication of progress towards an end, is founded, I think, upon what is very clearly a complete misinterpretation of the physical laws under which all are agreed that these actions take place.

2.—I shall endeavour to arrange what I have to say in two divisions, taking the quotation from Playfair, as it were, as the text :—First, The motions of the heavenly bodies ; the earth as one of them : and, Secondly, The phenomena presented by the earth's crust.

3.—Now, in the first place, the motions of the heavenly bodies are subject to resistance, which was not taken into account in the investigations of the French mathematicians. They gave out the theorem, that so far as the mutual attractions between the sun and the planets, and the law of inertia affecting the motions of each body, without any opposition of resistance, are concerned, certain disturbances known to exist among the motions of the heavenly bodies cannot become infinite, but must oscillate within certain limits.

4.—For instance, during a period—very many thousands of years say—the eccentricity of the earth's orbit round the sun may go on increasing. It might be supposed that that eccentricity could go on increasing so much, that at last the earth's path might cross that of one of the other planets. Serious disturbances in the motions of the two bodies might result, or even a fatal collision. But the theorem of the French mathematicians asserts, that while the eccentricity might go on increasing for a certain time, it has its limits ; thus declaring that there are oscillations and variations, but no continued variation in one direction. And this is a very important theorem undoubtedly. On details of the formula expressing it are founded all the calculations of modern physical astronomers regarding what are called the secular variations of the elements of the planetary orbits. But the French mathematicians were quite aware that, in making this statement, they neglected resistance. Those who quoted the grand theorem at which they arrived, did not perceive that exclusion. English philosophers and naturalists might surely have taken warning from Newton's simple brief decisive statement, " *majora autem planetarum et cometarum corpora motûs suos et progressivos et circulares, in spatiis minus resistentibus factos, conservant diutius ;* "[1] and have at least to some degree limited and qualified the expressions we so often meet in their popular writings, implying a perpetuity of the "existing order," past and future.

5.—Laplace was perfectly aware of the existence of resistance to fluid motion. In his theory of the tides, he points out most distinctly that if oscillation were established on the surface of the ocean—oscillation on a grand scale affecting the oceans—the waters of the Atlantic, for instance, swelling up, and those of the Pacific shrinking down, time about—that if such an oscillation were, by any force made to commence, then, in a very short time, he says " probably in a few months,"

---

[1] *Principia.* "Explanation of First Law of Motion."

we might expect it would altogether subside ; and in his theory of the tides, he treats the motion of the sea altogether as a motion of oscillation. There then is a tacit admission of the fact of resistance. But that tidal resistance influences the rotation of the earth, or, by reaction, the motions of the moon and sun, Laplace does not explicitly state. The modern theory of energy was imperfectly understood by Laplace and Lagrange. Lagrange, it is true, gave a foundation for the mathematical treatment of Dynamics, in which the theory of energy was the one great principle ; but he did not point out the application of the theory of energy to some of the consequences which now, in the present state of science, interest us perhaps more than any other conclusions which have been drawn from mathematical and physical reasoning. I am therefore entitled to speak so far of the science of energy as modern, although it was from Toricelli, Newton, John Bernouilli, and Lagrange, that we have learned the abstract dynamical principles of the science of energy. Even this abstract theory of energy teaches, that if there is resistance of any kind (against the tidal motion of the waters, for instance), *that* resistance must react upon some body, and take from that body, or from bodies connected with the phenomena, energy.

6.—The cause of the tides, as every one knows, of course, is the attraction of the moon and sun. The fact that the moon attracts the portion of the sea nearest to her more than she attracts the centre of the earth, and the centre of the earth than the remote parts of the ocean, gives rise to a tendency to draw water towards the moon, and leave a protuberance on the other side from the moon. That is the tendency; but the water of the ocean never gets time to take the exact form to which it tends. Just as if a large bath were suddenly tilted up and let down again : the water in it, at the time it was tilted up, tended to alter itself according to the new position of the bath, but there was no time for that tendency to have static effect ; so it is for the waters of the ocean when the moon comes to be over, for instance, the middle of the Atlantic, and *tends* to draw the water towards her then position and to leave it protuberant on the remote side of the earth. It is curious that in books of navigation the *tendency* has been so often spoken of as if that were the effect. An interesting correspondence occurred in the columns of the *North British Daily Mail* about a year ago, in which Newton's theory of the tides was disproved out of Norrie's *Navigation.* Norrie, in his work, describes the tendency; Newton in his theory describes the tendency, but points out that the waters of the ocean are in a state of continual oscillation and reverberation as it were (between two opposite continents, for instance, as between Africa on the one side and America on the other), and that at no one instant does the tendency have static effect according to what has been called " the equilibrium tide." Now it is the imaginary equilibrium tide that is often described as the theoretical tide in books on navigation, though the many readers of these books, with limited information as to what was written by Newton, Laplace, and Airy, accuse Newton of all the errors they have been taught.

7.—When we consider the moon as causing the tides, and the change from high to low as depending on the rotation of the earth, it becomes obvious that if there is resistance to the motion of the water that constitutes the tides, that resistance must directly affect the earth, and must react on those bodies, the moon and the sun, whose attractions cause the tides. The theory of energy declares, in perfectly general terms, that as there is frictional resistance, there must be loss of energy somewhere. We are not now merely content to say there is loss of energy by resistance, but the modern theory must account for what becomes of that energy. It is particularly to Joule that the full establishment of the true explanation as to what becomes of energy that is lost in friction is due. I suppose every one here present knows Joule's explanation, namely, that heat is generated. The friction of the waters against the bottom of the sea and against one another, in rubbing, so to speak, as they must to move about, to rise in one place and fall in another—the friction of waters especially

C 2

in the channels where there are tide races, gives rise to the generation of heat. Well, now, the end, where it altogether leaves our earth to be dissipated through space, is heat. The beginning to which we can at present trace the first source of that energy is in the motions of the moon and the earth. A little consideration shows us, by a very general kind of reasoning, that that particular component of the motion which at zero would give rise to no tides must tend to become zero. This we see as included in a very general proposition applicable to every possible case of action in nature. Now, if the motion of the earth in its rotation, relative to the moon in its revolution round the earth, were zero, there would be no rise and fall of water in lunar tides ; the earth would always turn the same face to the moon, and then it would be always high water towards the moon, low water in the intermediate circle, and high water from the moon, but there would be no motion of the waters relatively to the earth and so no friction. The tendency of friction must then, according to the general principle, be to reduce the relative motions of the earth and moon to that condition. However, it is satisfactory to know that we do not need to base a conclusion on so excessively general terms of the theory of energy as those. It is easy to see that the mutual action between the moon and the earth must tend, in virtue of the tides, to diminish the rapidity of the earth's rotation, and increase the moment of the moon's motion round the earth.

8.—" The tidal spheroid," you must understand, is not a reality, because the waters do not cover the whole earth, as we are here on *terra firma* to know. But there is a perfectly definite surface, being an elliptic spheroid calculated by mathematical rule, which is such that if it were the outer boundary of a distribution of water over a globe perfectly covered with water, this mass of water would exercise to an extremely close approximation the same force upon any distant particle of matter, and experience the same reacting force, as our tidally disturbed waters really do. That is what is properly called the tidal spheroid. It averages,

as it were, for the whole globe, the tidal effect
of the disturbing body considered. The tidal
spheroid averaging the moon's effect alone, is
called the luni-tidal spheroid ; and that for the
sun is called the soli-tidal spheroid. The resultant
tidal spheroid, representing on the same principle
the average displacement of the water produced by
the combined influence of the two bodies, is found
by simply adding the displacements from the undisturbed figure, represented respectively by the
luni-tidal and soli-tidal spheroids.

9.—If there were no frictional resistance against
the tides, each separate tidal spheroid would have
its longest diameter perpendicular to the line joining the centre of the earth with that of the disturbing body, whether moon or sun.[1] When the joint

[1] This assertion is founded not on observation, but on dynamical
principles. It depends on the truth that, if the tide-generating
influence of either sun or moon were suddenly to cease, the period
of the chief oscillation that would result would be *greater* than
either twelve solar or twelve lunar hours. The period of this
oscillation would be *less* than either twelve lunar or twelve solar
hours, if the sea were very much deeper than it is, or if it were
considerably deeper, and also less obstructed by land. (*See* § 11.)
If this were the case, the greatest axes of the luni-tidal and soli-
tidal spheroids would be in line with the moon and sun respectively,

influence of the sun and moon is analyzed by mathematical reasoning, it is found that there would be for either separately, a tidal spheroid fulfilling the condition just defined. Thus the dynamical result of the tendency of either body would be *low water* at the time of the high water of the imaginary equilibrium tide, and *vice versâ*, on the average of the whole earth. By the lunar tide, for instance, there would be low water when

and there would be average high water of either component tide when the body to which it is due crosses the meridian; also, the average times of greatest tide would still be those of new and full moon. But if the depth of the sea and the configuration of the land were such that the chief period of oscillation could be intermediate between twelve solar and twelve lunar hours, the greatest axis of the luni-tidal spheroid would be in line with the moon; but that of the soli-tidal spheroid would be perpendicular to the line joining the earth's and moon's centres. In this case, the times of spring tides would be those of quarter moons. In the first of these two unreal cases, the effect of tidal friction would be to make the time of average high water somewhat *later* in each component tide, than the time when the body producing it crosses the meridian; and this deviation would be greater for the sun than for the moon. Thus, the time of spring tides would be, as it is, somewhat later than the times of new moon and full moon. But, in the second of the imagined cases, the effect of friction would be to advance the time of solar high water and to retard the time of lunar high water; and thus the time of spring tides would be somewhat before the times of the quarter moons.

the moon is crossing the meridian, and (supposing for simplicity, the moon to be in the plane of the earth's equator) there would be high water when she is rising and setting. When the moon and sun are exactly in conjunction and opposition, the longest and shortest axes of their tidal spheroids would agree ; and the highest and lowest tides on the average of the whole earth would be the high water and low water immediately before and after, or after and before, the time of new and full moon. This, be it remembered, is on the supposition of no tidal resistance, but does not involve any assumption whatever of regularity, whether as to the boundary of the sea or as to uniformity of its depth.

10.—Now, it is well known that, in this part of the world, the "spring tides" are observed to be late by from a day and a half to three days after new moon and full moon. On the West Coast of Ireland the interval is about thirty-six hours ; it amounts to about sixty hours at London Bridge, and has intermediate values at intermediate points of the British Channel.

Along the Atlantic Coast of Europe the interval seems to be between eighteen hours, which may be about its amount at the Cape of Good Hope, and thirty-six hours its value on the West Coast of Ireland; and it is probable that in all seas the spring tides are *late* by an interval of, in general, something more than twelve hours and less than three days after the time of new moon and full moon. Hence, the crests of the luni-tidal and soli-tidal spheroids are not coincident when the earth, moon, and sun are in one line, but are coincident at some time, probably exceeding twelve hours, after the moon has crossed the line joining the earth and sun. This then is decisive in showing a sensible effect of resistance to the tidal motions, as was first, I believe, remarked by Airy.[1] That there must be such resistance is quite certain to us, from our knowledge of the properties of matter; but it is very interesting, and it is very important with reference to the subject of my present statement, to find a sensible effect on the average tides

[1] "Tides and Waves," § 544—*Encyclopedia Metropolitana.*

of the whole ocean due to resistance against the tidal motions.

11.—The accompanying woodcut illustrates the position of the sun and moon, and of the longest axis of the resultant tidal spheroid at the average time of spring tides for the whole earth. It represents a section in the plane of the equator, in which for simplicity the sun and moon are both supposed to be. The spectator being supposed to look at the diagram from the north side, sees the earth rotating, and the moon revolving round the earth's centre, each in the direction opposite to the motion of the hands of a watch. If there were no tidal friction, OMS would be in one straight line, and HH′, the longest axis of the tidal spheroid would be perpendicular to it. What observation on the time of spring tides proves, is simply that OM is inclined to OS forwards by the angle through which the moon advances before the sun, in the time by which the spring tides are late. If this time were twelve hours, the angle MOS would be 6°. The dynamical theory

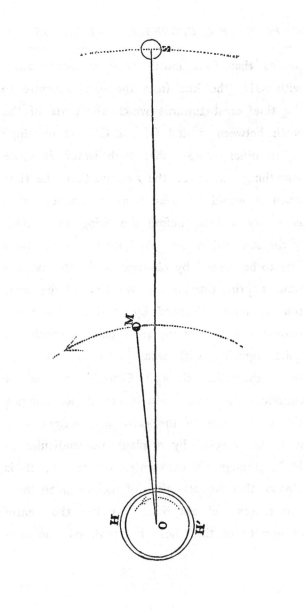

proves that each must make an acute angle with OH, the line from the earth's centre to the tidal crest towards which the parts of the earth between it and OM or OS are rotating; or, in other words, that high water is made something *earlier* on the average than the time when it would be were there no friction; that is to say, a little before the rising and setting of the sun and moon. And thus what we have seen to be proved by observation on the average time of spring tides, is that the time of the lunar tide is more advanced by frictional resistance than is the time of the solar tide : a conclusion quite agreeing with what is to be anticipated by mathematical theory.[1] Considering now for simplicity the lunar tide alone, if we imagine the whole mass of the earth and waters upon it to be bisected by a plane perpendicular to HH′, through O, the centre of gravity, it is obvious that the attractions of the moon on these two halves will not balance round the centre of gravity of the whole, but that, on the con-

[1] *See* footnote on § 9.

trary, the combined influence of the greater attraction on the nearer protuberance of the waters (H), and of the less attraction on the more remote protuberance (H') would, if the whole were rigid, tend to turn the line H'H towards the direction OM. If the round earth rotated inside the waters, these not sharing its rotation, the effect would be as if a mechanical friction strap or belt were applied round the earth's equator, and were held from turning by a couple of forces equal in moment (or rotational importance) to that calculated as what is technically called "the moment round the earth's axis" of the moon's attraction on the two protuberances.

12.—But the waters do not get pulled back by the moon as a whole. They are, as a whole, drawn along with the solid earth by friction on the bottom, and by friction of water on water. Therefore, from century to century, the water moves along with the earth. Though it is due, in the first place, to a force in the water, the resultant effect on the earth and the water is the same as if the whole were a solid globe rotating

inside the supposed friction-strap. The amount
of each of these forces constituting the supposed
couple holding back the equatorial friction-strap,
would be equal to the weight of two million tons,
according to the hypothesis and calculations taken
from the Rede Lecture (Cambridge, May, 1866)
on the "dissipation of energy," in the abstract
appended to the present article. This hypothesis
supposes HH' to be inclined to OM, at an angle
of 45°, the position in which, with a given amount
of protuberance, the tidal retardation of the
earth's rotation would be a maximum: having
been assumed for the purpose of estimating a
superior limit to the conceivable amount of the
influence in question. The resulting retardation
would be the same as if the earth had (as a
common "terrestrial globe,") pivots at the North
and South poles, each half an inch diameter, and
resisting forces were applied tangentially to these
pivots, amounting in all to four thousand million
million tons force. With the same supposed
degree of protuberance in the luni-tidal spheroid,[1]

[1] This is such as would make the average amount of rise of tide

the frictional resistance would be something between the amount estimated, and one-fifth of it, if the angle HOM were anything between 45° and 84°. But if, as is most probably the case, the average lateness of the spring tides behind the full and change, amounts at least to twelve hours, the angle MOS cannot be less than 6°, this being the angle through which the moon moves in her orbit in twelve hours; and HOS is certainly something short of 90° Hence it is almost certain that HOM has some value between 45° and 84°. I conclude that either the average spheroidal tide must be less than $1\frac{1}{3}$ ft. or the amount of resistance to the earth's rotation must exceed one-fifth of the amount which I had estimated as a superior limit; the only doubtful assumption being, that the lateness of the spring tides behind the times of full and change, is not less on the average than twelve hours.

13.—The general tendency of that action, then, is to diminish the velocity of the earth's rotation from lowest to highest, for the whole surface of the earth, $1\frac{1}{3}$ feet, if the tides were everywhere, those of the luni-tidal spheroid covering the whole earth.

round its axis, and lengthen the duration of the day. That there is such a tendency has long been known to philosophers of the more abstract kind. It is difficult to say who first promulgated this idea. It has been recently stated that the metaphysician Kant first asserted that the earth's rotation is diminished through the influence of the tides. This I know for certain, that the idea was first given to me by my brother, Professor James Thomson. As long ago as the first meeting of the British Association in Glasgow, 1840, he propounded as a necessary result of the theory of energy, that friction of the tides in channels must give rise to a loss of something then called *vis-viva*, from the motion of the earth and the moon. More recently published articles, and especially those of Mayer, the great German advocate of the modern theory of heat, who did so much to urge the reception of the idea of an equivalence between heat and mechanical power, point out that the rotation of the earth must be diminished by the tides.

14.—But we may go further, and say that tidal

action on the earth disturbs, by re-action, the moon. The tidal deformation of the water exercises the same influence on the moon as if she were attracted not precisely in the line towards the earth's centre, but in a line slanting very slightly, relatively to her motion, in the direction forwards. The moon, then, continually experiences a force forward in her orbit by re-action from the waters of the sea. Now, it might be supposed for a moment that a force acting forwards would quicken the moon's motion ; but, on the contrary, the action of that force is to retard her motion. It is a curious fact easily explained, that a force continually acting forward with the moon's motion will tend, in the long run, to make the moon's motion slower, and increase her distance from the earth. On the other hand, the effect of a resisting force on, for instance, the earth would undoubtedly be, in the course of ages, to make the earth go faster and faster round the sun. The reason is, that the resistance allows the earth to fall in a spiral path towards the sun, whose attraction generates more velocity than frictional resistance destroys. The tidal deformation of the water on the earth tends, on the whole, therefore, to retard the moon's angular motion in her orbit; but (by the accompanying augmentation of her distance from the earth) to increase the *moment of her motion* round the earth's centre and the ultimate tendency—so far as the earth's rotation is concerned—must be to make the earth keep always the same face to the moon.

15.—It may be remarked, in passing, that the corresponding tendency has probably already had effect on the moon itself. The moon always turns the same face to the earth. If the moon were now a liquid mass, there would be enormous tides in it. The friction in that fluid would cause the moon to tend to turn the same face towards the earth: and we find the moon turns the same face always to the earth. It seems almost inevitable to our minds, constituted as they are to connect possible cause and real effect, and say that a possible cause is a real cause; and thus to believe the reason why the moon turns always the same side to us is because it was once a liquid mass which experienced tides and viscous resistance against the tidal motion. The only other view we can have —the only other hypothesis we can make—is that the moon was created with such an angular velocity as to turn always the same face to the earth. But the course of speculative and physical science is absolutely irresistible as regards the relation between cause and effect. Whenever we can find a possible antecedent condition of matter, we cannot help inferring that that possible antecedent did really exist as a preceding condition— a condition, it may be, preceding any historical information we can have—but preceding and being a condition from which the present condition of things has originated by force acting according to laws controlling all matter. The theory that the moon has been brought to her present condition of rotation by tidal friction of her own mass was, I believe, first given by Helmholtz. I cannot say so certainly, because so many philosophers have speculated and drawn conclusions regarding antecedents of the solar system from very general philosophic principles; but, so

far as I know, that view was first given by
him.

16.—It is impossible, with the imperfect data we
possess as to the tides, to calculate how much their
effect in diminishing the earth's rotation really is.
But even from such data as those referred to in
§ 11, it can be shown that the tidal retardation
of the earth's rotation must be something very
sensible. Still, it is unsatisfactory to be in the
position of asserting that we know there must be
a retardation (we cannot tell how much) and then
to be told, in opposition to that theory, that
observations of ancient eclipses make it certain
that the length of the day has not varied by one
ten-millionth part of twenty-four hours from 721
years before the Christian era.[1] The calculation
was first made by Laplace. It depended in part
on the historical facts of two eclipses of the moon,
seen in Babylon, one of them March 19, 721 B.C.,
which was first perceived " when one hour after
her rising was fully past ; "[2] the other on the

---

[1] Poisson *Traité de Mécanique*, Sec. 433, Vol. ii., Ed. 1833.
[2] Dunthorne " On the Acceleration of the Moon," *Phil. Trans*
1749. (Hutton's *Abridgement*, Vol. ix.)

22nd December, 313 years B.C., which was first perceived "half an hour before the end of night," and which, though now known to have lasted only about an hour and a half, had not come to an end when the moon set. The rotation of the earth cannot have experienced *much* retardation during these 2700 years, or else the moon-rise must have taken place *after* instead of *before* the beginning of the first of those eclipses; and it cannot have experienced much acceleration,[1] or else the moon must have set at Babylon before the second mentioned eclipse commenced, which, therefore could not have been seen from that place. But Dunthorne showed that these records and various observations regarding many other less ancient eclipses all agree in *demonstrating* the correctness of a suspicion which Halley had raised

---

[1] As regards the earth's rotation, it seems to have been only *acceleration* (due to cooling and shrinking) that was suspected until Kant and others showed that the tides must produce retardation. But Laplace proved, by calculations founded on Fourier's theory of the conduction of heat (not at all on astronomical data), that the acceleration by shrinking on account of cooling cannot have shortened the day by as much as 1-300th of a second of time; that is to say, by about a twenty-five millionth of its own amount in the last 2,000 years.

that the moon's mean angular motion has been accelerated somewhat *relatively to the earth as time-keeper ;* and he estimated the amount of this acceleration to be 20 seconds of angular velocity per century gained per century.[1] Laplace, accepting this conclusion, attempted to explain it by showing that the planets cause indirectly an acceleration of the moon's angular velocity through their influence in producing a secular diminution of the eccentricity of the earth's orbit. The principle is admitted, and to Laplace is attributed, and must always be attributed, the very great discovery of the cause of an apparent secular acceleration of the moon's mean motion. He calculated out the results of this discovery, and they seemed to tally precisely with the supposition that the earth's velocity of rotation had been constant since 721 B.C. ;[2] but in 1853 our great English physical astronomer, Adams, pointed out

[1] Grant's *History of Physical Astronomy*, p. 60.

[2] Laplace's theory gave him 21″·7 per century of angular velocity as the average gain per century during the last 2,500 years, which exceeds Dunthorne's estimate from observation by 8½ per cent. — Grant's *History of Physical Astronomy*, p. 63.

an error of a technical kind in Laplace's process—
the omitting to take into account *in the tangential
component* of the sun's disturbing force on the
moon, the disturbing influence of the variation of
the eccentricity of the earth's orbit, and he worked
out the theory with this correction.

17.—The result, roughly stated, was to halve
the amount of acceleration calculated by Laplace
and to leave half of Dunthorne's observed *relative*
acceleration of the moon to be accounted for
otherwise. In 1853, Adams communicated to
Delaunay, one of the great French mathematicians,
his final result, that at the end of a century the
moon is 5·7 seconds of angle in advance of the
position she had when relatively to the meridian
of the earth, according to the angular velocity of
the moon's motion at the beginning of the century,
and the acceleration of the moon's motion truly
calculated from the various disturbing causes then
recognised. This, then, shows an unaccounted for
gain per century of 11″·4 per century of angular
velocity of the moon's motion, on the hypothesis
that the earth's angular velocity is uniform.

Delaunay soon after verified this result, and about the beginning of 1866 suggested that the true explanation may be the retardation of the earth's rotation by tidal friction. Using the hypothesis that the cause of the discrepancy is retardation by tidal friction, and allowing for the consequent retardation of the moon's mean motion, Adams, in an estimate which he has recently worked out in conjunction with Professor Tait and myself, found, on a certain assumption as to the proportion of retardations due to the moon and the sun—that 22 seconds of time is the error by which the earth would in a century get behind a thoroughly perfect clock rated at the beginning of the century.

18.—Thus the most probable result that physical astronomy gives us up to the present time · is that the earth is not an accurate chronometer, but, on the contrary, is getting slower and slower, if tested by a truly perfect clock—a clock as good as an astronomical clock ought just now to be, and that is at least 200 times as good as astronomical clocks are—because astronomical clocks are just as great a disgrace to the mechanical genius of Europe and America as chronometer watches are a credit. Astronomical clocks go only about two or three times as well as pocket chronometer watches ; although the latter, from the continual agitations to which they are exposed, are in very disadvantageous circumstances. When they shall be made two or three hundred times as good as they are, we shall have an instrument which, for use during a few centuries, will be a superior time-keeper to the earth ; and it will not then be necessary to set the clock by the stars, but we shall test the earth's motion by the clock. However, that is only in anticipation. Perhaps we may not live to see that use of the clock. In the meantime we are obliged to put up with the earth and stars as a means for regulating our clocks. Failing a good clock to check the earth by, we have to take the best we can find and apply corrections to it. The moon is a very unequal time-keeper, but by prodigious labour, carried out by Newton, Clairaut, Laplace, Plana, Hansen, Adams, and Delaunay, the errors in the moon's motion are very accurately known. The moon's rotation round the earth is as it were a clock hand going round in about 29 days, and the earth is as it were the hand of another clock going round in 24 hours. The only timekeeper by which we can at present test the accuracy of the earth's motion is the moon. Imperfect as the moon is, an error has, you see, been discovered in the earth as a time-keeper, on reference to the moon. Consider that fact, and see whether it justifies the statement I have referred to by Playfair in his Illustrations of Hutton's theory, that there is nothing in the motions of the heavenly bodies that tends to their own dissolution or to a permanent alteration of the existing state of things. For instance, no resistance tending to stop the progress of the earth!

19.—Now, if the earth is losing angular velocity at that great rate, at what rate might it have been rotating a thousand million years ago? It must have been rotating faster by one-seventh part than at present, and the centrifugal force must have been greater in the ratio of the square of 8 to the square of 7, that is, in the ratio of 64 to 49. There must have then been more centrifugal force at the equator due to rotation than now in the proportion of 64 to 49. What does the theory of geologists say to that? There is just now at the equator one two-hundred-and-eighty-ninth part of the force of gravity relieved by centrifugal force. If the earth rotated seventeen times faster bodies would fly off at the equator. The present figure of the earth agrees closely with the supposition of its having been all fluid not many million years ago.

20.—The centrifugal force a hundred million years ago would be greater by about 3 per cent. than it is now, according to the preceding estimate of tidal retardation; and nothing we know regarding the figure of the earth, and the disposition of land and water, would justify us in saying that a body consolidated when there was more centrifugal force by 3 per cent. than now might not now be in all respects like the earth, so far as we know it at present. But if you go back to ten thousand million years ago—which does

not satisfy some great geologists—the earth must
have been rotating more than twice as fast
as at present—and if it had been solid then, it
must be now something totally different from what
it is. Now, here is direct opposition between
physical astronomy, and modern geology as represented by a very large, very influential, and,
I may also add, in many respects, philosophical
and sound body of geological investigators, constituting perhaps a majority of British geologists.
It is quite certain that a great mistake has been
made—that British popular geology at the present
time is in direct opposition to the principles of
natural philosophy. Without going into details
I may say it is no matter whether the earth's
lost time is 22 seconds, or considerably more or
less than 22 seconds, in a century, the principle is
the same. There cannot be uniformity. The
earth is filled with evidences that it has not been
going on for ever in the present state, and that
there is a progress of events towards a state infinitely different from the present.

21.—But it is not only to the effect of the

tides that we refer for such conclusions. Go to other bodies besides the earth and moon ; consider the sun. We depend on the sun very much for the existing order of things. Life on this earth would not be possible without the sun, that is, life under the present conditions—life such as we know and can reason about. When Playfair spoke of the planetary bodies as being perpetual in their motion, did it not occur to him to ask, what about the sun's heat ? Is the sun a miraculous body ordered to give out heat and to shine for ever ? Perhaps the sun was so created. He would be a rash man who would say it was not— all things are possible to Creative Power. But we know also, that Creative Power has created in our minds a wish to investigate and a capacity for investigating ; and there is nothing too rash, there is nothing audacious, in questioning human assumptions regarding Creative Power. Have we reason to believe Creative Power did order the sun to go on, and shine, and give out heat for ever ? Are we to suppose that the sun is a perpetual miracle ? I use the word *miracle* in the sense

of a perpetual violation of those laws of action
between matter and matter which we are allowed
to investigate here at the surface of the earth,
in our laboratories and mechanical workshops.
The geologists who have uncompromisingly
adopted Playfair's maxim have reasoned as if
the sun were so created. I believe it was
altogether thoughtlessness that led them ever to
put themselves in that position ; because these
same geologists are very strenuous in insisting
that we must consider the laws observable in
the present state of things as perennial laws. I
think we may even consider them as having gone
too far in assuming that we must consider present
laws—a very small part of which we have been
able to observe—as sufficient samples of the
perennial laws regulating the whole universe in
all time. But I believe it has been altogether an
oversight by which they have been led to neglect
so greatly the fact of the sun's heat and light.

22.—The mutual actions and motions of the
heavenly bodies have been regarded as if light had
been seen and heat felt without any evolution of

mechanical energy at all. Yet what an amount of
mechanical energy is emitted from the sun every
year! If we calculate the exact mechanical value
of the heat he emits in 81 days we find it
equivalent to the whole motion of the earth in her
orbit round the sun. The motion of the earth in
her orbit round the sun has a certain mechanical
value ; a certain quantity of steam power would be
required, acting for a certain time, to set a body
as great as the earth into motion with the same
velocity. That same amount of steam power
employed for the same time in rubbing two stones
together would generate an enormous quantity of
heat, as much heat as the sun emits in 81 days.
But suppose the earth's motion were destroyed,
what would become of the earth? Suppose it
were to be suddenly, by an obstacle, stopped in its
motion round the sun? It would suddenly give
out 81 times as much heat as the sun gives out in
a day, and would begin falling towards the sun,
and would acquire on the way such a velocity that,
in the collision, a blaze of light and heat would be
produced in the course of a few minutes equal to

what the sun emits in 95 years. That is, indeed, a prodigious amount of heat; but just consider the result if all the planetary bodies were to fall into the sun. Take Jupiter with its enormous mass, which, if falling into the sun, would in a few moments cause an evolution of 32,240 years' heat. Take them all together—suppose all the planets were falling into the sun—the whole emission of heat due to all the planets striking the sun, with the velocities they would acquire in falling from their present distances, would amount to something under 46,000 years' heat. We do not know these figures very well. They may be wrong by ten or twenty or thirty per cent., but that does not influence much the kind of inference we draw from them. Now, what a drop in the ocean is the amount of energy of the motion of the planets, and work to be done in them before they reach their haven of rest, the sun, compared with what the sun has emitted already! I suppose all geologists admit that the sun has shone more than 46,000 years? Indeed, all consider it well established that the sun has already, in geological periods, emitted ten, twenty, a hundred, perhaps a thousand—I won't say a hundred thousand—but perhaps a thousand times as much heat as would be produced by all the planets falling together into the sun. And yet Playfair and his followers have totally disregarded this prodigious dissipation of energy. He speaks of the existing state of things as if it must or could have been perennial.

23.—Now, if the sun is not created a miraculous body, to shine on and give out heat for ever, we must suppose it to be a body subject to the laws of matter (I do not say there may not be laws which we have not discovered), but, at all events, not violating any laws we have discovered or believe we have discovered. We must deal with the sun as we should with any large mass of molten iron, or silicon, or sodium. We do not know whether there is most of the iron, or the silicon, or the sodium—certainly there is sodium ; as I learned from Stokes before the end of the year 1851 ; and certainly, as Kirchhoff has splendidly proved, there is iron. But we must reason upon the sun as if it were some body having properties such as bodies we know have. And this is also worthy of attention :—naturalists affirm that every body the earth has ever met in its course through the universe, has, when examined, been proved to contain only known elements—chemical substances such as we know and have previously met on the earth's surface. If we could get from the sun a piece of its substance cooled, we should find it to consist of stone or slag, or metal, or crystallised rock, or something that would not astonish us. So we must reason on the sun according to properties of matter known to us here.

24.—In 1854, I advocated the hypothesis that the energy continually emitted as light (or radiant heat) might be replenished constantly by meteors falling into the sun from year to year ; but very strong reasons have induced me to leave that part of the theory then advocated by me which asserted that the energy radiating out from year to year is supplied from year to year ; and to adopt Helmholtz's theory, that the sun's heat was generated in ancient times by the work of mutual gravity between masses falling together to form his body. The strongest reason which compelled me to give up the former hypothesis was, that the amount of bodies circulating round the sun within a short distance of his surface, which would be required to give even two or three thousand years of heat, must be so great, that a comet shooting in to near the sun's surface and coming away again, would inevitably show signs of resistance to a degree that no comet has shown. In fact, we have strong reason to believe that there is not circulating round the sun, at present, enough of meteors to constitute a few thousand years of future sun-heat. If, then, we are obliged to give up every source of supply from without—and I say it advisedly, because there is no sub-marine wire, no "underground railway," leading into the sun—we see all round the sun, and we know that there is no other access of energy into the sun than meteors,—if, then, we have strong reason to believe that there is no continual supply of energy to the sun, we are driven to the conclusion that it is losing energy. Now, let us take any reasonable view we can.

E 2

Suppose it is a great burning mass, a great mass
of material not yet combined, but ready to combine,
a great mass of gun-cotton, a great mass of gunpowder, or nitro-glycerine, or some other body
having in small compass the potential elements
of a vast development of energy.   We may imagine
that to be the case, and that he is continually
burning from the combustion of elements within
himself ; or we may imagine the sun to be merely
a heated body cooling ; but imagine it as we please,
we cannot estimate more on any probable hypothesis, than a few million years of heat.   When I
say a few millions, I must say at the same time,
that I consider one hundred millions as being a
few, and I cannot see a *decided* reason against
admitting that the sun may have had in it one
hundred million years of heat, according to its
present rate of emission, in the shape of energy.
An article, by myself, published in *Macmillan's
Magazine* for March, 1862, on the age of the sun's
heat,[1] explains results of investigation into various
questions as to possibilities regarding the amount

[1] Republished as Appendix (E) to Thomson and Tait's *Natural
Philosophy*, Vol i., Part ii., 2nd ed. (1885).

of heat that the sun could have, dealing with it as you would with a stone, or a piece of matter, only taking into account the sun's dimensions, which showed it to be possible, that the sun may have already illuminated the earth for as many as one hundred million years, but at the same time also rendered it almost certain that he had not illuminated the earth for five hundred millions of years. The estimates here are necessarily very vague, but yet vague as they are, I do not know that it is possible, upon any reasonable estimate, founded on known properties of matter, to say that we can believe the sun has really illuminated the earth for five hundred million years.

25.—But Playfair looks to the earth, and says that while the heavenly bodies give every evidence of having gone on for ever as now, the earth, in the phenomena presented all through its crust, to unprejudiced observers, gives similar evidence, and seems to indicate no evidence of a beginning, and no progress or advance towards an end. Now, let us consider the question of underground heat. The earth, if we bore into it anywhere, is warm, and if we could apply the test deep enough, we should, no doubt, find it very warm. Suppose you should have here before you a globe of sandstone, and boring into it found it warm, boring into another place found it warm, and so on, would it be reasonable to say that that globe of sandstone has been just as it is for a thousand days? You would say, "No; that sandstone has been in the fire, and heated not many hours ago." It would be just as reasonable to take a hot water jar, such as is used in carriages, and say that that bottle has been as it is for ever—as it was for Playfair to assert that the earth could have been for ever as it is now, and that it shows no traces of a beginning, no progress towards an end.

26.—There have been feeble attempts to reason away the argument from under-ground heat. The geologists, to whose theory I object, do at the same time, I believe, admit that the temperature increases downwards, wherever observations have been made. They have hitherto taken a somewhat supine view of the subject. Admitting that there is in many places evidence of an increase of temperature downwards, they say they have not evidence enough to show that there is increase of temperature downwards in all parts of the earth, or enough of evidence to allow us to say that the theory that accounts for underground heat, by local chemical action, may not be true. This being the state of the case as regards underground heat, where must we apply to get evidence? Observation; observation only. We must go and look. We must bore the earth here in the neighbourhood. We must examine underground temperature in other places. We must send out and bore under the African deserts, where water has not reached for hundreds of years. The whole earth must be made subject to a geothermic survey. Having been deeply impressed with these views for many years, I have long endeavoured, in vain, to call the attention of geologists to them. I now feel very greatly indebted to the Geological Society of Glasgow, for giving me the opportunity of speaking of them this evening. I may be allowed to add that on the occasion of the recent meeting of the British Association, at Dundee, the import-

ance of investigation of underground temperature
was not denied by geologists, before whom the
subject was brought in the first instance, on that
occasion, by a paper by the Hungarian naturalist,
Schwarez. A result of the discussion which
followed the reading of that paper was the appoint-
ment of a committee [1] for investigating underground
temperature.

27.—The laws of the progress of summer heat
and winter cold downwards were investigated
thoroughly by the great French mathematician
Fourier, and made the subject of observation in
different localities. We know very well now what
temperature, so far as the annual variation is concerned, may be expected to be found at ten,
twenty, or thirty feet down, according to the conductivity and capacity for heat of the strata. If
we bore down to a depth of twenty-four feet we may
find, in mid-winter, the highest temperature. Prob-

[1] [Note of Jan. 4, 1893.] This Committee has been reappointed
from year to year ever since, and, under the persevering and able
guidance of Prof. Everett, continues to obtain and publish valuable
information regarding underground temperature in all parts of the
world.

ably, last midsummer's heat is now about reaching thirty feet below this place. Principal Forbes instituted experiments on the Calton Hill, at the Craigleith Quarry, and the Experimental Gardens, and in these three places the observations were continued for several years, the temperature being observed every week. From these observations, he calculated the conducting powers of the different strata, and his results were, I believe, the first obtained of an accurate kind, regarding the conducting power of rock in its natural condition in the earth's crust. Ångström made similar experiments in Sweden, and deduced results on the same principles. Similar observations were made at Greenwich, and calculated by Dr. Everett; so from these results we may consider the conductivity of ordinary surface rocks as generally very well known.

28.—But the question, how much does temperature increase downwards from hundred feet to hundred feet, is one which has been but very imperfectly investigated indeed. Observation of temperature in mines, as Schwarez points out, and as Phillips pointed out in the Geological Society of London, are very unsatisfactory. Air circulating through the mines, and water percolating and being pumped out, give rise to disturbances so great, that we cannot say if in a lower level of a mine, we find a colder temperature than in a higher level, the result is due to colder strata. The best ventilated deep mine will be the coolest ; and in passing, I may remark, which is, perhaps, of some interest in the present and prospective state of the question of the supply of coal, that we know no limit of the depth to which coal may be worked, depending on terrestrial temperature. Suppose there was coal, or rather charcoal, where the strata were red hot, it might be gone into and that with perfect ease. All that is necessary is plenty of ventilation. This will keep the temperature cool enough for working, and thus there is no limit whatever to the depth to which the miner may proceed. I do not say it would not be enormously more expensive to bring up coal (gas-coke, or charcoal) from four thousand fathoms, if there is any at so great a depth, than to bring up what we call coal from a depth of one hundred or two hundred fathoms, but that it could be got at and brought out, notwithstanding even a red hot temperature of the surrounding strata is quite certain. Plenty of ventilation, conducted on proper thermodynamic principles,[1] will give quite a satisfactory temperature for the workers in the mines.

[1] That is to say, the air must be compressed and cooled at the upper surface, or at some convenient place in the mine or shaft, at no great depth below the surface. This cold dense air must be conducted to the lowest levels through a strong enough pipe, and allowed to expand into the mine through an engine, or engines, like a common high-pressure engine working expansively. A great part of the work of this engine must be spent otherwise than in generating heat in the mine ; for instance, it may be used for working the gear to raise the mineral. A portion of its work may be spent in cutting out the minerals, as is sometimes done at present by compressed-air engines ; but it must be remembered that the full dynamical equivalent of this part of the work of the engine is developed in heat in the mine. Probably best practical plan for working very deep mines will be to employ the engine power used at the surface all in compressing air ; the compressed air to be cooled, either by water, if there is a sufficient cold water supply, or by radiation to the sky, and by atmospheric convection. This condensed air should be used for working the engine or engines in proper places at the great depths required to work the gear for raising the minerals, etc., and small cutting engines in various parts of the workings. Thus a sufficient supply of cool air may be distributed through the mine. If the ordinary method of ventilation by *drawing out* air, whether by an air pump, or by a fire burning at

29.—All sound naturalists agree that we cannot derive accurate knowledge of underground temperature from mines. But every bore that is made for the purpose of testing minerals gives an opportunity of observation. If a bore is made, and is left for two or three days, it will take the temperature of the surrounding strata. Let down a thermometer into it, take proper means for ascertaining its indications, draw it up, and you have the measure of the temperature at each depth. There are most abundant opportunities for geothermic surveys in this locality by the numerous bores made with a view to testing minerals, and which have been left either for a time or permanently without being made the centre of a shaft. Through the kindness of Mr. Campbell, of Blyths-

the foot of the vertical shaft be used, the down current of fresh air will be warmed to the amount of nearly 1° C. for every fifty fathoms of descent, by the natural compression of the air through its own weight or more exactly 18° cent. per 1000 fathoms ; being $\frac{1}{180}$ of a degree centigrade per foot, according to an investigation which I have given in the *Proceedings of the Manchester Literary and Philosophical Society*, January 1862, "On the Convective Equilibrium of Temperature in the Atmosphere." (*Mathematical and Physical Papers*, Vol. iii., p. 255.)

wood, several bores in the neighbourhood of his house have been put at the disposal of the committee of the British Association, to which I have referred. In one of these bores very accurate observations have been made, showing an increase of temperature downwards, but which is not exactly the same in all the strata, the difference being, no doubt, due to different thermal conductivities of their different substances. I need not specify minutely the numbers, but I may say in a general way, that the average increase is almost exactly $\frac{1}{50}$ of a degree Fahrenheit per foot of descent; which agrees with the estimate generally admitted as a rough average for the rate of increase of underground temperature in other localities.

Another bore has been put at the disposal of the committee, and the investigation of it is to be commenced immediately, so that I hope in the course of a few days some accurate results will be got. It has been selected because the mining engineer states in his report that the coal has been very much burned or charred, showing the effect of heat; and it becomes an interesting question, Are there any remains of that heat that charred the coal in ancient times; or has it passed off so long ago that the strata are now not sensibly warmer on account of it?

30.—I shall conclude by simply referring to calculations regarding the quantity of heat at present conducted out from the interior of the earth, which I have given in an article, entitled "The 'Doctrine of Uniformity' in Geology, briefly refuted;"[1] and to analytical investigations regarding *antecedents* of the present condition of underground heat contained in a paper "On the Secular Cooling of the Earth,"[2] appended to the volume on *Natural Philosophy* by Professor Tait and myself, recently published. The first of these shows, by mere calculation of the actual conduction, that the present rate of increase of underground temperature could not last for twenty or thirty thousand million years, without there being dissipated out of the earth as much heat as would be given off by a quantity of ordinary surface rock equal to 100 times the earth's mass, cooling from 100° cent.

[1] *Proceedings R. S. E.*, December, 1865, and p. 6, above.
[2] First republished in the *Trans. R. S. E.*, 1862, published in *Mathematical and Physical Papers*, Vol. iii., p. 295.

to 0°. In the second, by the analytical investigation of antecedents it is shown that the present condition implies either a heating of the surface, within the last 20,000 years of as much as 100 degrees, Fahr., or a greater heating all over the surface at some time farther back than 20,000 years.[3]

Now, are geologists prepared to admit that at some time within the last 20,000 years there has been all over the earth so high a temperature as that? I presume not; no geologist—no *modern* geologist—would for a moment admit the hypothesis that the present state of underground heat is due to a heating of the surface at so late a period as 20,000 years ago. If that is not admitted we are driven to a greater heat at some time more than 20,000 years ago. A greater heating all over the surface than 100 degrees Fahr., would kill nearly all existing plants and animals, I may safely say. Are geologists prepared to say that all life was killed off the earth 50,000, 100,000, or 200,000 years ago? For the uniformity theory, the farther back the time of high surface tempera-

---

[1] Thomson and Tait, Appendix D. § (j.)

ture is put the better ; but the farther back the
time of the heating, the hotter it must have been.
The best for those who draw most largely on time
is that which puts it farthest back, and that is the
theory that the heating was enough to melt the
whole.   But even if it was enough to melt the
whole, we must still admit some limit, such as fifty
million years, one hundred million years, or two or
three hundred million years ago.[1]   Beyond that we
cannot go.   The argument described (§ 19) above
regarding the earth's rotation shows that the earth
has not gone on as at present for a thousand
million years.   Dynamical theory of the sun's heat
renders it almost impossible that the earth's surface has been illuminated by the sun many times
ten million years.   And when finally we consider
underground temperature we find ourselves driven
to the conclusion in every way, that the existing
state of things on the earth, life on the earth, all
geological history showing continuity of life, must
be limited within some such period of past time as
one hundred million years.

[1] Thomson and Tait, Appendix D. § (r.)

APPENDIX.

ON THE OBSERVATIONS AND CALCULATIONS REQUIRED TO FIND THE TIDAL RETARDATION OF THE EARTH'S ROTATION.[1]

THE first *publication* of any definite estimate of the possible amount of the diminution of rotatory velocity experienced by the earth through tidal friction is due, I believe, to Kant. It is founded on calculating the moment round the earth's centre of the attraction of the moon, on a regular spheroidal shell of water symmetrical about its longest axis, this being (through the influence of fluid friction) kept in a position inclined backwards at an acute angle to the line from the earth's centre to the moon. One of the simplest ways of seeing the result is this :—First, by the known conclusions as to the attractions of ellipsoids, or still more easily by the consideration of the proper "spherical harmonic "[2] (or Laplace's coefficient) of the second degree, we see that an equipotential surface lying close to the bounding

[1] From the Rede Lecture, Cambridge, May 23, 1866, "On the Dissipation of Energy."

[2] Thomson and Tait's *Natural Philosophy*, § 536 (4).

F

surface of a nearly spherical homogeneous solid ellipsoid is approximately an ellipsoid with axes differing from one another by three-fifths of the amounts of the differences of the corresponding axes of the ellipsoidal boundary. Now it is known [1] that a homogeneous prolate spheroid of revolution attracts points outside it approximately as if its mass were collected in a uniform bar having its ends in the foci of the equipotential spheroid. If, for example, a globe of water of 21,000,000 feet radius (this being nearly enough the earth's radius) be altered into a prolate spheroid with longest radii exceeding the shortest radii by two feet, the equipotential spheroid will have longest and shortest radii differing by $\frac{9}{5}$ of a foot. The foci of this latter will be at 7,100 feet on each side of the centre ; and therefore the resultant of gravitation between the supposed spheroid of water and external bodies will be the same as if its whole mass were collected in a uniform bar of 14,200 feet length. But by a well-known proposition,[2] a uniform line FF′ (a diagram is unnecessary) attracts a point M in the line M K bisecting the angle F M F′ Let C Q be a perpendicular from C, the middle point of F′F, to this bisecting line M K. If C M be $60 \times 21 \times 10^6$ (the moon's distance), and if the angle F C M be 45° we find, by elementary geometry, CQ=·02 of a foot (about

---

[1] Thomson and Tait's *Natural Philosophy*, § 501 and § 480 (*e*).
[2] *Ibid.*, § 480 (*b*) and (*a*).

$\frac{1}{4}$ inch). The mass of a globe of water equal in bulk to the earth is $\cdot 97 \times 10^{21}$ tons.[1] And, the moon's mass being about $\frac{1}{80}$ of the earth's, the attraction of the moon on a ton at the earth's distance is $\frac{1}{80} \times \frac{1}{60^2}$, or $\frac{1}{290,000}$ of a ton force, if, for brevity, we call a ton force the ordinary terrestrial weight of a ton—that is to say, the amount of the earth's attraction on a ton at its surface. Hence the whole force of the moon on a globe of water equal in bulk to the earth is $\frac{\cdot 97 \times 10^{21}}{290,000}$, or $3 \cdot 3 \times 10^{15}$ tons force. If, then, the tidal disturbance were exactly what we have supposed, or if it were (however irregular) such as to have the same resultant effect, the retarding influence of the moon's attraction would be that of $3 \cdot 3 \times 10^{15}$ tons force acting in the plane of the equator and in a line passing the centre at $\frac{1}{50}$ of a foot distance. Or it would be the same as a simple frictional resistance (as of a friction-brake) consisting of $3 \cdot 3 \times 10^{15}$ tons force acting tangentially against the motion of a pivot or axle of about $\frac{1}{2}$ inch diameter. To estimate the retardation produced by this, we shall suppose the square of the earth's radius of gyration,

[1] In stating large masses, if English measures are used at all, the ton is convenient, beeause it is 1000 kilogrammes nearly enough for many practical purposes and rough estimates. It is 1016·047 kilogrammes ; so that a ton diminished by about 1·6 per cent. would be just 1000 kilogrammes.

instead of being $\frac{2}{5}$, as it would be if the mass were homogeneous, to be $\frac{1}{3}$ of the square of the radius of figure, as it is made to be, by Laplace's probable law of the increasing density inwards, and by the amount of precession calculated on the supposition that the earth is quite rigid. Hence (if we take $g=32\cdot2$ feet per second generated per second, and the earth's mass $=5\cdot3 \times 10^{21}$ tons) the loss of angular velocity per second, on the other suppositions we have made, will be

$$\frac{32\cdot2 \times 3\cdot3 \times 10^{15} \times \cdot02}{5\cdot3 \times 10^{21} \times \frac{1}{3} (21 \times 10^{6})^{2}} \text{, or } 2\cdot7 \times 10^{-21}.$$

The loss of angular velocity in a century would be $31\frac{1}{2} \times 10^{8}$ times this, or $8\cdot5 \times 10^{-12}$, which is as much as $\frac{1\cdot16}{10^{7}}$ of $\frac{2\pi}{86400}$, the present angular velocity. Thus in a century the earth would be rotating so much slower that, regarded as a time-keeper, it would lose about $1\cdot16$ seconds in ten million, or $3\cdot6$ seconds in a year. And the accumulation of effect of uniform retardation at that rate would throw the earth as a time-keeper behind a perfect chronometer (set to agree with it in rate and absolute indication at any time) by 180 seconds at the end of a century, 720 seconds at the end of two centuries, and so on. In the present very imperfect state of clock-making (which scarcely produces an astronomical clock two or three times more accurate than a marine chronometer or good pocket-watch) the only chronometer by which we can check the earth is one which goes much worse —the moon. The marvellous skill and vast labour devoted to the lunar theory by the great physical astronomers Adams and Delaunay, seem to have settled that the earth has really lost in a century about ten seconds of time on the moon corrected for all the perturbations which they had taken into account. M. Delaunay has suggested that the true cause may be tidal friction, which he has proved to be probably sufficient by some such estimate as the preceding.[1] But the many disturbing influences to which the earth is exposed render it a very untrustworthy time-keeper. For instance, let us suppose ice to melt from the polar regions (20° round each pole, we may say) to the extent of something more than a foot thick, enough to give 1·1 foot of water over those areas, or ·066 of a foot of water if spread over the whole globe, which would in reality raise the sea-level by only some such almost undiscoverable difference as $\frac{3}{4}$ of an inch, or an inch. This, or the reverse, which

[1] It seems hopeless, without waiting for some centuries, to arrive at any approach to an exact determination of the amount of the actual retardation ef the earth's rotation by tidal friction, except by extensive and accurate observation of the amounts and times of the tides on the shores of continents and islands in all seas, and much assistance from *true* dynamical theory to estimate these elements all over the sea. But supposing them known for every part of the sea, the retardation of the earth's rotation could be calculated by quadratures.

we may believe might happen any year, and could certainly not be detected without far more accurate observations and calculations for the mean sea-level than any hitherto made, would slacken or quicken the earth's rate as a time-keeper by one-tenth of a second per year.[1]

Again an excellent suggestion, supported by calculations which show it to be not improbable, has been made to the French Academy by M. Dufour, that the retardation of the earth's rotation indicated by M. Delaunay, or some considerable part of it, may be due to an increase of its moment of inertia by the incorporation of meteors falling on its surface. If we suppose the previous average moment of momentum of the meteors round the earth's axis to be zero, their influence

---

[1] The calculation is simply this. Let E be the earth's whole mass, a its radius, k its radius of gyration before, and k' after the supposed melting of the ice, and W the mass of ice melted. Then, since $\frac{2}{3}a2$ is the square of the radius of gyration of the thin shell of water supposed spread uniformly over the whole surface, and that of either ice-cap is very approximately $\frac{1}{2}a^2$ (sin 20°)^2, we have

$$E k'^2 = E k^2 + W a^2 \left[\tfrac{2}{3} - \tfrac{1}{2}(\sin 20°) \right].$$

And by the principle of the conservation of moments of momentum, the rotatory velocity of the earth will vary inversely as the square of its radius of gyration. To put this into numbers, we take, as above, $k^1 = \frac{1}{3}a^2$ and $a = 21 \times 10^6$. And as the mean density of the earth is about $5\frac{1}{2}$ times that of water, and the bulk of a globe is the area of its surface into $\frac{1}{3}$ of its radius,

$$\mathrm{E} : \mathrm{W} : : \frac{55a}{3} : \text{·066}$$

will be calculated just as I have calculated that of the supposed melting of ice. Thus meteors falling on the earth in fine powder (as is in all probability the lot of the greater number that enter the earth's atmosphere and do not escape into external space again) enough to form a layer about $\frac{1}{20}$ of a foot thick in 100 years, if of 2·4 times the density of water, would produce the supposed retardation of 10ˢ on the time shown by the earth's rotation. But this would also accelerate the moon's mean motion by the same proportional amount; and therefore a layer of meteor-dust accumulating at the rate of $\frac{1}{40}$ of a foot per century, or 1 foot in 4,000 years, would suffice to explain Adams and Delaunay's result. I see no other way of directly testing the probable truth of M. Dufour's very interesting hypothesis than to chemically analyze quantities of natural dust taken from any suitable localities (such dust, for instance, as has accumulated in two or three thousand years to depths of many feet over Egyptian, Greek, and Roman monuments). Should a considerable amount of iron with a large proportion of nickel be found or not found, strong evidence for or against the meteoric origin of a sensible part of the dust would be afforded.

Another source of error in the earth as a time-keeper, which has often been discussed, is its shrinking by cooling. But I find by the estimates

I have given elsewhere [1] of the present state of deep underground temperatures, and by taking $\frac{1}{100000}$ as the vertical contraction per degree centigrade of cooling in the earth's crust, that the gain of time on this account by the earth, regarded as a clock, must be extremely small, and may even not amount to more in a century than $\frac{1}{30}$ of a second or $\frac{1}{6000}$ of the amount estimated above as conceivably due to tidal friction.

[1] "Secular Cooling of the Earth," *Transactions of the Royal Society of Edinburgh*, 1862; and *Philosophical Magazine*, January, 1863.

OF GEOLOGICAL DYNAMICS

[Address to the Geological Society of Glasgow,
April 5, 1869.]

PART   I.—*Reply to Professor Huxley's Address to the*
Geological Society of London, of February 19,
1869.
PART II.—*Origin and Total Amount of Plutonic Energy.*
PART III.—*Note on the Meteoric Theory of the Sun's Heat.*

PART I.

1. IN a recent address (February 19th, 1869) to
the Geological Society of London, from the Presidential Chair, Professor Huxley directs attention
to the two following sentences, which he quotes
from my lecture on "Geological Time," delivered
to this Society on the 27th February, 1868 :—

" A great reform in geological speculation seems
" now to have become necessary. . . . . It is quite

"certain that a great mistake has been made—that
" British popular geology, at the present time, is
'in direct opposition to the principles of natural
"philosophy."

2. Professor Huxley attempts to answer these
charges, and appeals to "that higher court of
"educated scientific opinion to which we are all
"amenable," for a verdict of "not guilty." He
prefaces "his pleading" with the following remarkable statement :—"As your attorney-general for
" the time being, I thought I could not do better
"than get up the case with a view of advising you.
" It is true that the charges brought forward by
" the other side involve the consideration of matters
"quite foreign to the pursuits with which I am
"ordinarily occupied ; but in that respect I am
" only in a position which is, nine times out of ten,
" occupied by counsel, who, nevertheless, contrive
" to gain their causes, mainly by force of mother-
" wit and common sense, aided by some training
" in other intellectual exercises."

I must, therefore, in the beginning, be permitted
to say that the very root of the evil to which I

object is that so many geologists are contented to regard the general principles of natural philosophy, and their application to terrestrial physics, as matters quite foreign to their ordinary pursuits. I must also say, that though a clever counsel may, by force of mother-wit and common sense, aided by his very peculiar intellectual training, readily carry a jury with him to either side, when a scientific question is before the court, or may even succeed in perplexing the mind of a judge ; I do not think that the high court of educated scientific opinion will ever be satisfied by pleadings conducted on such precedents. But jury and judge may be somewhat perplexed as to what it is on which they are asked to give verdict and sentence, when they learn that Professor Huxley himself makes the gravest of the accusations against Hutton and Uniformity which he repels as made by me. In the course of his address he describes Kant's Cosmogony ; and pointing out anticipations in it of some of the "great principles" taught in the *Theory of the Earth*, somewhat later, by Hutton, he says, "On

" the other hand, Kant is true to science. He
" knows no bounds to geological speculation, but
" those of intellect. He reasons back to a begin-
" ning of the present state of things ; he admits the
" possibility of an end." Professor Huxley does
not use words without a meaning : and these mean
that Hutton was *not* true to science, when he said,
" The result, therefore, of this physical inquiry is,
" that we find no vestige of a beginning, no pros-
" pect of an end." The chief complaint on which
I am now brought into court is, that I have
extended the same accusation to modern followers
of Hutton who have used this dictum as a fundamental maxim of their geology.

3. In opening his case, Professor Huxley asks,
" What is it to which Sir W. Thomson refers when
" he speaks of ' geological speculation ' and ' British
" ' Popular Geology ' ? " then enters on a highly
interesting and instructive discussion of various
schools of geological philosophy, which constitutes
the chief substance of his address, and recurs to
the question, " Which of these is it that Sir William
" Thomson calls upon us to reform ? " But instead

of answering the question he says, " It is obviously
" Uniformitarianism " which Sir W. Thomson
" takes to be the representative of geological
" speculation in general." I have given no ground
for this statement. Not merely " obviously," but
avowedly and explicitly, I attacked Uniformitarianism ; but I did not attack geological speculation in general. On the contrary, I anxiously
and carefully guarded every expression of my
complaint from applicability to other speculations
than those involving more or less fundamentally
the particular fallacies against which my objections
were directed ; and the very phrases I used to limit
my accusations showed that I had not taken
Uniformitarianism to be the representative of
geological speculation in general. The geology
which I learned thirty years ago in the University
of Glasgow embodied the fundamental theory now
described and approved by Professor Huxley as
Evolutionism. This I have always considered to
be the substantial and irrefragable part of geological speculation ; and I have looked on the
ultra-uniformitarianism of the last twenty years

as a temporary aberration worthy of being ener-
getically protested against.

4. In the course of his lecture, Professor Huxley
says :—" I do not suppose that at the present day
" any geologist would be found to maintain
" absolute uniformitarianism, to deny that the
" rapidity of the rotation of the earth *may* be
" diminishing, that the sun *may* be waxing dim,
" or that the earth itself *may* be cooling. Most of
" us, I suspect, are Gallios, 'who care for none of
" ' these things,' being of opinion that, true or
" fictitious, they have made no practical difference
" to the earth, during the period of which a record
" is preserved in stratified deposits."

It is precisely because so many geologists " have
cared for none of these things," which (though not
matters of words merely) do certainly belong to
the law of Nature, that they have brought so much
of British popular geology into direct opposition
to the principles of Natural Philosophy. Professor
Huxley tells us that they have been of opinion
that the secular cooling of the earth has made
no practical difference to it during the period of

which a record is preserved in stratified deposits.
On what calculation is this opinion founded?
One considerable part of the reform in geological
speculation for which I ask is, that evidence
adduced in favour of the opposite opinion should
be thoroughly sifted, and not merely disposed of
as matters of opinion, or of faith beyond the realm
of reason.

5. It was, however, in reference to the special
subject of my paper, " Geological Time," that I
chiefly urged the necessity of reform, and it is
satisfactory now to see that in this respect
considerable progress must have been made, when,
on the 19th February, 1869, Professor Huxley
ventured before the Geological Society of London
to suggest that " the limitation of the period
" during which living beings have inhabited this
" planet to one, two, or three hundred million
" years, may be admitted, without a complete
" revolution in geological speculation." When he
says that on me rests the *onus probandi* of my
assertion in January, 1868, " that a great reform
seemed to have become necessary," as I had

brought "forward not a shadow of evidence" in support of that assertion, I cannot complain that he puts a heavy burden on me. No moderately well read or well instructed student of modern British popular geology wants evidence from me, in addition to that supplied by his reminiscences of books and lectures, that the admission of such a limit as even worthy of attention, is a sweeping reform. Here, however, is some of it, if desired. [The italics are mine in each case.]

6. "So[1] that, in all probability, a far longer "period than 300 million years has elapsed *since* "*the latter part of the secondary period.*"

7. "Again,[2] where the FORCE seems unequal "to the result, the student should never lose sight "of the element TIME : *an element to which we* "*can set no bounds in the past,* any more than we "know of its limit in the future.

"It will be seen from this hasty indication that "there are two great schools of geological "causation—the one ascribing every result to the

---

[1] Darwin's *Origin of Species*, Edition 1859, page 287.

[2] Page's *Advanced Text Book of Geology*, 1859, page 338.

"ordinary operations of Nature, combined with the
"element of *unlimited time*, the other appealing to
"agents that operated during the earlier epochs
"of the world with greater intensity, and also for
"the most part over wider areas. *The former*
"*belief is certainly more in accordance with the*
"*spirit of right philosophy*, though it must be
"confessed that many problems in geology seem
"to find their solution only through the admission
"of the latter hypothesis."

8. "Any[1] person who has paid even the
"slightest attention to the science of geology must
"be aware of the fact that the whole of our
"knowledge in regard to age in this science is
"confined to relative age, and that with respect to
"absolute age we have little or no real informa-
"tion ; and in this absence of positive knowledge
"as to the absolute age of rocks, geologists have
"sometimes indulged in the wildest and most
"extraordinary statements and speculations. *They*
"*speak of the enormous lapse of time requisite for the*
"*formation of exceedingly small quantities of rock,*

[1] *Manual of Geology.* By the Rev. S. Haughton, F.R.S. Edition
1865, p. 79.

"*in a manner that would almost make us suppose*
"*that some miraculous agency was at work to retard*
"*the progress of the formation of these rocks.* Indeed
"it has been well observed that the mantle of the
"preachers has fallen on the geologists, and that
"the figures and images by which the former paint
"to their terrified audience the duration of
"eternity, *a parte post* have been seized on, and
"adopted by the geologists in endeavouring to
"describe eternity *a parte ante.* The infinite time
"of the geologists is in the past ; and *most of their*
"*speculations regarding this subject seem to imply*
"*the absolute infinity of this time,* as if the human
"imagination was unable to grasp the period of
"time requisite for the formation of a few inches
"of sand or feet of mud, and its subsequent consoli-
"dation into rock."

"Professor Thomson[1] has made an attempt to
"calculate the length of time during which the
"sun can have gone on burning at the present rate,
"and has come to the following conclusion :—'It
"'seems, therefore, on the whole, most probable

[1] *Manual of Geology.* By the Rev. S. Haughton, F.R.S. Edition 1865, p. 82.

" 'that the sun has not illuminated the earth for
" ' 100,000,000 years, and almost certain that he
" 'has not done so for 500,000,000 years. As for
" ' the future, we may say with equal certainty that
" 'the inhabitants of the earth cannot continue
" ' to enjoy the light and heat essential to their
" ' life for many million years longer, unless new
" ' sources now unknown to us, are prepared in the
" ' great storehouse of creation.' "

" *This result* of Professor Thomson's, *although*
" *very liberal in the allowance of time, has offended*
" *geologists, because, having been accustomed to deal*
" *with time as an infinite quantity at their disposal,*
" *they feel naturally embarrassment and alarm at*
" *any attempt of the science of Physics to place a*
" *limit upon their speculations.* It is quite possible
" that even a hundred million of years may be
" greatly in excess of the actual time during which
" the sun's heat has remained constant."

" Although [1] I have spoken somewhat disrespect-
" fully of the geological calculus in my lecture, *yet*

[1] *Manual of Geology.* By the Rev. S. Haughton, F.R.S. Edition
1865, p. 99.

" *I believe that the time during which organic life*
" *has existed on the earth is practically infinite,*
" *because it can be shown to be so great as to be*
" *inconceivable by beings of our limited intelligence.*"

9. " The [1] only agent to which we can reasonably
"attribute the destruction and removal of masses
" of rock, notwithstanding that they were many
"thousands of feet in thickness, and many hundred
"thousand square miles in extent, is the slow and
" gradual gnawing of the sea breakers upon coasts,
" an action always tending to plane down land to a
" little below the level of the upper surface of the
" ocean."

" The time required for such a slow process to
" effect such enormous results *must of course*
" *be taken to be inconceivably great.* The word
" 'inconceivably' is not here used in a vague, but
" in a literal sense, to indicate that the lapse
" of time required for the denudation that has
" produced the present surfaces of some of the
" older rocks, is vast beyond any idea of time

---

[1] Students' *Manual of Geology.* By J. B. Jukes, M.A., F.R.S.,
1862.

"which the human mind is capable of con-
" ceiving.

" Mr. Darwin, in his admirably-reasoned book
" on the origin of species, so full of information
" and suggestion on all geological subjects,
" estimates the time required for the denudation
" of the rocks of the weald of Kent, or the erosion
" of space between the ranges of chalk hills, known
" as the North and South Downs, at *three hundred*
" *millions of years.*[1] The grounds for forming this
" estimate are of course of the vaguest description.
" It may be possible, perhaps, that the estimate
" is a hundred times too great, and that the real
" time elapsed did not exceed three million years ;
" but, on the other hand, *it is just as likely that the*
" *time which actually elapsed since the first com-*
" *mencement of the erosion till it was nearly as*
" *complete as it now is, was really a hundred times*

[1] Prof. Phillips refers to this estimate of Mr. Darwin's ; prefers
one inch per annum to one inch per century as the rate of erosion ;
and says that most observers would consider even the one inch per
annum too small for all but the most invincible coasts ! He thus,
on purely geological grounds, reduces Mr. Darwin's estimate of the
time to less than one one-hundredth.—PHILLIPS's *Life on the Earth.*
Cambridge, 1860 (Rede Lecture).

"*greater than his estimate, or thirty thousand* "*millions of years.*"

10. It is to be presumed that Professor Huxley repudiates these figures when he says, "If we " accept the limitation of time placed before us " by Sir William Thomson, it is not obvious on " the face of the matter that we shall have to alter " or reform our ways in any appreciable degree : " but I am at a loss to understand how he can ask, ' Has it ever been denied that this period *may* be enough for the purpose of geology ? "

11. In marked contrast to them, is Professor Phillips's careful analysis of "the geological scale of time."[1] By reckoning the actual thicknesses of different strata, and allowing $\frac{1}{111}$ of an inch per annum as a not improbable mean rate at which they have been deposited, he finds ninety-six million years as a possible estimate for the anti- quity of the base of the stratified rocks ; but he gives reasons for supposing that this may be an over estimate, and finds that from stratigraphical

[1] Phillips's *Life on the Earth.* Cambridge, 1860 (Rede Lecture), p. 119.

evidence alone, we may regard the antiquity of life
on the earth as being possibly between thirty-eight
millions and ninety-six millions of years. How
many orthodox geologists accepted these estimates
fourteen months ago? Now, indeed, we have a
precisely similar estimate from Professor Huxley
himself. And just twelve months ago at a meeting
of this Society, Mr. Geikie,[1] declaring his secession
from the prevailing orthodoxy, maintained that all
the erosion of which we have monumental evidence
in stratified rocks, and in the shapes of hills and
valleys over the world, could have taken place
several times over in the period of a hundred
million years.

12. Professor Huxley, immediately after his
statement (quoted in § 10 above), " If we accept
" the limitation of time placed before us by Sir
" William Thomson, it is not obvious on the face
" of the matter that we shall have to alter or
" reform our ways in any appreciable degree," says
" We may therefore proceed with much calmness
" and, indeed, *much indifference to the result*, to

[1] Now Sir Archibald Geikie (Note of date January 9, 1893).

" enquire whether that limitation is justified by
" the arguments employed in its support." [The
italics are mine.] This method of treating my
" case " is perfectly fair, according to the judicial precedents upon which Professor Huxley professedly
founds his pleading. I make no comment or reply,
but simply ask permission to put in the following
evidence [the italics again are mine] :—" He who
" can read Sir Charles Lyell's grand work on the
" Principles of Geology, which the future historian
" will recognise as having produced a revolution in
" natural science, yet does not admit how *incom-*
"*prehensibly vast* have been the past periods of
" time, *may at once close this volume.*" (Darwin's
Origin of Species by means of Natural Selection.)[1]

13. In the discussion in this Society which
followed my lecture on " Geological Time," the
necessity for much longer periods in geological
history than 100 million years was very strongly
urged on biological grounds. I answered that
Geologists, by estimates of very great numbers
of millions of years, had misled biologists into

[1] Edition 1859 ; page 282.

hypotheses which could not now be justly adduced
to support such estimates when physical geology
declares against them. I am glad to find this view
supported by the high authority of Professor
Huxley himself, who says, " But it may be said
" that it is biology and not geology which asks for
" so much time—that the succession of life demands
" vast intervals ; but this appears to me to be
" reasoning in a circle. Biology takes her time
" from geology. The only reason we have for
" believing in the slow rate of the change in
" living forms is the fact that they persist through
" a series of deposits which geology informs us
" have taken a long while to make. If the geo-
" logical clock is wrong, all the naturalist will have
" to do is to modify his notions of the rapidity of
" change accordingly." But I may be permitted
to remark that a correction of this kind cannot be
said to be unimportant in reference to biological
speculation. The limitation of geological periods,
imposed by physical science, cannot, of course,
disprove the hypothesis of transmutation of species ;
but it does seem sufficient to disprove the doctrine

that transmutation has taken place through " descent with modification by natural selection."

14. And now as to Professor Huxley's examination of my arguments. (I.) Referring to my estimate of the retardation of the earth's rotational velocity due to an imagined melting of ice from the polar regions, he remarks that a certain accumulation of polar ice since the miocene epoch, and not more than he imagines may in reality have taken place, would produce five times as much acceleration as the amount of the retardation which we have estimated from the tides ; and he supposes that this would "leave $\frac{4}{5}$ of a second per annum in the way of acceleration." But *the observed result is retardation*, and Professor Huxley's hypothesis as to ice, if it were valid, would therefore prove retardation by the tides six times as much as that which we have ventured to estimate ! I am much obliged to him for this suggestion, and also to Mr. Croll for a suggestion which he has made to me that the erosion of equatorial mountains and deposition of detached matter at considerable distances from the equator, in either north or south

latitude, may be exerting, at the present time, an accelerating influence of a sensible amount upon the earth's rotational velocity, and rendering the observed retardation less than that due to the tides. For, as shown in my lecture on *Geological Time* (§12 and Appendix), the dynamical theory of the tides, and known facts regarding the interval between "full and change of the moon," and the times of spring tides, render it difficult to see how tidal retardation of the earth's rotation can be so little as to make the integral of lost time in a century amount to only twenty-two seconds. It is conceivable that something of this accumulation of ice suggested by Professor Huxley, or erosion of matter from the equator suggested by Mr. Croll, may, to a considerable extent, have temporarily counteracted the tidal retardation.

15. Now Professor Huxley asks, "If tidal re-"tardation can be thus checked and overthrown "by other[1] temporary conditions, what becomes "of the confident assertion based upon the

[1] I presume the presence of the word "other" here is to be regarded as an undetected "erratum."

" assumed uniformity of tidal retardation, that ten
" thousand million years ago the earth must have
" been rotating more than twice as fast as at present,
" and, therefore, that we geologists are ' in direct
" 'opposition to the principles of natural philos-
" 'ophy' if we spread geological history over that
" time ?" I answer that tidal retardation cannot
be permanently overthrown by temporary conditions ; that its true amount may be considerably
greater than that which we have estimated from
the theory of the moon's motion ; and that from
million of years to million of years it must always
be a positive retardation : whereas the integral
effect of the others in millions of years must be
zero.   Professor Huxley's remarks, instead of making my assertion less worthy of confidence, give
us a probability that we may repeat it with equal
confidence, *for a smaller limit than ten thousand
million years,* when in the course of a few years the
committee of the British Association on tides
gives us, for all seas, more knowledge of the times
of spring tides relatively to the changes of the
moon ; of the times of daily high water relatively

to the moon's transits; and of the amount of rise and fall, than we have at present.

16. But since Professor Huxley has raised the definite question—What interchange of water and ice would keep the rotation of the earth constant from the miocene period? I must point out that it can be answered only when we know how many centuries have elapsed, supposing we assume (as he does with me, for the sake of argument), a uniform datum of tidal retardation; and must remark that he has omitted to multiply his estimated thickness of ice by this unknown number of centuries. The subject is certainly somewhat perplexing, owing to the ambiguity of the words commonly used in expressing such matters; of which we have a familiar instance in the statement, " clock too fast," or " clock too slow," meaning *clock before*, or *clock behind.* Our estimate of tidal retardation is such as to make the earth, regarded as a clock, come to be twenty-two seconds of time behind at the end of the century,[1]

---

1 Or 25^2 times 22^s, that is, 13750^s, or $3^h 49^m 10^s$, at the end of 25 centuries.

after just beginning at the beginning of the century to go slow, and going gradually slower and slower, at a uniform rate of retardation during the century. Thus to get behind by twenty-two seconds at the end of the century implies going slower by 22 of a second per annum at the middle of the century and ·44 of a second per annum, at the end, than at the beginning of the century. This, therefore, gives a retardation of ·44 of a second per annum per century, or of ·0044 of a second per annum per annum ; an effect equal in amount to what would be produced by the melting of ·044 of a foot of ice per annum from ice caps of twenty degrees round each pole. Thus to produce an amount of retardation equal to that which we estimate as due to the tides, ice must melt at the rate of 044 of a foot per annum, or 4·4 feet per century from the polar ice caps.[1]   But

---

[1] The attraction of the polar ice upon the ocean referred to by M. Adhemar and Mr. Croll, was not taken into account in my calculations in the Rede Lecture of 1866, from which these figures are quoted. Its effect is to render a somewhat less thickness of ice, but greater depression of water in the equatorial regions, necessary to produce the same increase of rotational velocity.

if the actual retardation were not due to the tides its amount would be ten instead of twenty-two, by observation and dynamical theory of the moon's motion. Two feet of ice per century, therefore, melted from the supposed polar ice caps would be required to account for it by the melting of ice, or fifty feet in the twenty-five centuries during which it has taken place. If, then, Professor Huxley can show that it is probable that ice to any such extent as *that* has melted from polar regions, giving a gradual rise of the average level of the sea to the extent of three feet, in the last twenty-five centuries, he would establish the probability of another solution than tidal retardation to the astronomical question put before us by Adams. But the very fact that dynamical theory of the tides leads me to look for rather a greater than a less amount of retardation than the twenty-two seconds which we have estimated, makes it probable that no such considerable rising of the sea level, if any rising at all, will be found to have taken place. On this question we may, however, fairly look for some positive evidence from the investigations of geologists and archæologists combined.

17. My expectations from tidal dynamics now weigh with me very decidedly against M. Dufour's meteoric hypothesis ;—much more than they did at the time I first referred to it in the Rede Lecture of 1866. And although the establishment of this hypothesis would be almost as fatal as the retardation by tides to the uniformitarian geologists, I cannot view the solution of the question with indifference. I look forward with much interest to see it tested by chemical analysis of the dust which has accumulated over Egyptian, Greek, and Assyrian monuments for the last two or three thousand years.

18. (II.) The only answer which Professor Huxley gives to my argument from the sun's heat is, that as lately as fifteen years ago I "entertained " a totally different view of the origin of the sun's " heat, and believed that the energy radiated from " year to year was supplied from year to year, a " doctrine which would have suited Hutton per- " fectly." So far from this being the case, if Pro-

fessor Huxley will " Hansardize " me by looking to my original paper on " The Mechanical Energies of the Solar System," he will see that my contribution to the meteoric theory of solar heat was to prove the insufficiency of any chemical theory, and to point out that *meteoric supply cannot be perennial in even approximate uniformity with the existing order of things.*[1] I think he will find nothing in that paper which " justly entitles " him to " disregard " my present estimates, but, on the contrary, much to enforce them. In a note to that paper, dated May 4th, 1854, is to be found an indication of my subsequent correction of the untenable part of my first views, and, obstructing it, a difficulty which I then felt as to the sun's capacity for heat. In my article on the " Age of the Sun's Heat,"[2] to which Professor Huxley refers, a resolution of that difficulty is pointed out, according to which it is shown that the sun's capacity for heat is probably more than ten times, and less than 10,000 times

[1] Farther information on this point is to be found in an extract from the *Proceedings of the Glasgow Philosophical Society*, March 24, 1859, appended (Part III., below, p. 127).

[2] *Macmillan's Magazine*, March, 1862.

that of an equal mass of water under ordinary pressure. A British jury could not, I think, be easily persuaded to disregard my present estimate by being told that I have learned something in fifteen years.

19. (III.) Referring to my third line of argument founded on a consideration of terrestrial temperature, Professor Huxley asks the question, " But is " the earth nothing but a cooling mass, ' like a " hot-water jar, such as is used in carriages,' or " ' a globe of sand-stone,' and has its cooling been " uniform ? " and says, " An affirmative answer to " both these questions seems to be necessary to " the validity of the calculations on which Sir " W. Thomson lays so much stress." I reply that I have carefully considered the first question, and referred to it in my paper on the " Secular Cooling of the Earth," § 9,[1] or Thomson and Tait's *Natural Philosophy*, Appendix D, § i. ; and that the main purport of that paper constitutes a *negative answer to the second question.* I have distinguished the

[1] " Secular Cooling," § 18 ; *Transactions of the Royal Society of Edinburgh*, 1862 ; *Phil. Mag.*, 1862 ; or, Thomson and Tait, Appendix, D, § 5.

results calculated from conduction at only the present rate, giving a limit of twenty or thirty thousand million years, in a short article (p. 6 above) of more recent date entitled, " The Doctrine of Uniformity in Geology Briefly Refuted," from those of the analytical investigation of the " antecedents " of the present condition of underground heat, contained in my former paper (" Secular Cooling "). The analytical investigation shows the law of the *greater* rate of conduction outwards in past times, and demonstrates a much closer limit for the whole time during which the earth has been solid and continuously cool enough at its surface to be habitable without break of continuity to life, than can be estimated without taking into account the *deviation from uniformity* which I assert.

20. Referring partly to my views and partly to his own inadvertent misstatement of them Professor Huxley continues :—

" Nevertheless it may be urged that such affirma-
" tive answers are purely hypothetical, and that
" other suppositions have an equal right to consid-

H

" eration. For example, is it not possible that the
" prodigious [1] temperature which would seem to
" exist at 100 miles below the surface, all the
" metallic bases may behave as mercury does at a
" red heat, when it refuses to combine with oxy-
" gen ; while, nearer the surface, and therefore at a
" lower temperature, they may enter into combina-
" tion (as mercury does with oxygen a few degrees
" below its boiling point) and so give rise to a heat
" which is totally distinct [2] from that which they
" possess as cooling bodies ?   And has it not also
" been proved by recent researches that the quality
" of the atmosphere may immensely affect its per-
" meability to heat, and consequently profoundly
" modify the rate of cooling of the globe as a
" whole ?

" I do not think it can be denied that such
" conditions may exist and may so greatly affect
" the supply and the loss of terrestrial heat as to

[1] Does this imply internal fluidity? If so, it is to be rejected.
"Prodigious" seems too strong a word for any temperature below
the melting point of the material.

[2] By no means so : but, on the contrary, an essential part of the
heat emitted by the composite mass in cooling.

" destroy the value of any calculations which leave
" them out of sight."

I reply that I admit the first, and emphatically
deny the second, proposition of the last sentence.
Heat of combination of elements, present together
in a mixed mass and devoid of chemical affinity at
a high temperature, but acquiring chemical affinity
and consequently combining as the temperature
sinks, constitutes merely an addition to the sum of
the thermal capacities of the several elements
separately reckoned, to give the effective thermal
capacity of the composite mass. And the value
of " calculations " which leave this possibility " out
of sight " is not " destroyed " though an altered
figure in the result might be necessitated by an
altered estimate of specific heat. But in my calculations I have left a wide enough margin to give
due weight on Professor Huxley's side to the
smallness of our knowledge regarding specific
heats, thermal conductivities, and temperatures of
fusion, of the earth's material. And as to the
cloudiness or clearness of the atmosphere, I say
that the secular cooling of the earth is not affected

by it.   The one question relevant tc atmospheric effect on the secular cooling of the earth is, what has been the resulting temperature of the upper surface of land and sea ?   My calculations depend only on the assumption that through geological history this temperature has been suitable for such life as now exists on the earth.

21.   Criticising the calculations I had adduced regarding the earth's rotation, Professor Huxley makes the following remarks, which have equal bearing upon those regarding the sun's heat and light and the earth's interior temperature : " I desire " to point out that this seems to be one of the " many cases in which the admitted accuracy of " mathematical processes is allowed to throw a " wholly inadmissible appearance of authority over " the results obtained by them.   Mathematics may " be compared to a mill of exquisite workmanship " which grinds you stuff of any degree of fineness ; " but, nevertheless, what you get out depends on " what you put in ; and as the grandest mill in the " world will not extract wheat flour from peascods, " so pages of formulæ will not get a definite result

" out of loose data." To the second of these sentences I assent, but certainly not to the first. I have not presented definite results; I have amply indicated how "loose" my data are; and I have taken care to make my results looser. Professor Huxley himself in *other* parts of his address has *complained of their vagueness* "as greatly embarrassing the discussion." If I had presumed to limit the past duration of life on the earth to one million years or to ten million years, by calculations, founded on such data as I have used, so ill drawn an inference could scarcely "embarrass" those who are still disposed to trust to "a practically unlimited bank " of time ready to discount any amount of hypo- " thetical paper." But it is obvious that they must be seriously embarrassed by even a superior limit of four hundred million years: especially when the declaration of it is coupled with the assertion of a *very strong probability* that "all geological history showing continuity of life," is in reality to be condensed into a period not exceeding *one* hundred million years.

22. Before concluding, I may be permitted to

make a few remarks on the practical bearing of the limitations which I have adduced upon some points of geological theory, which, when the boundary between mineralogy and geology is once passed, cannot be evaded even by those most averse to speculation.

23. Fourier's theory of the conduction of heat renders it almost impossible to escape the conclusion, that if the earth has been solid and habitable continuously during the last fifty million years, its rate of increase of underground temperature per metre downwards must have been very sensibly more rapid fifty million years ago than now. The more recently discovered laws of thermodynamics render it certain that the sun must have been something very different fifty million years ago from what he is now; and almost certain that he must have been then very much hotter. And we find Sir Roderick Murchison[1] writing as follows, on purely stratigraphical grounds:—" I could here " cite the works of many eminent writers for " numerous evidences of the grander intensity of

---

[1] *Siluria;* 1867 Edition, page 489.

" causation in former epochs, by which gigantic
" stratified masses were sometimes inverted, or so
" wrenched, broken, and twisted, as to pass under
" the very rocks out of which they were formed
" Among those who have passed away l may
" mention de Saussure ; Von Buch, Humboldt,
" Cuvier, Brongniart, Buckland, Conybeare, De la
" Becke, and W. Hopkins. Of those who hold
" the same views, and are now living, I may
" enumerate Elie de Beaumont, D'Archiac, De
" Verneuil, Studer, Sedgwick, J. Forbes, Phillips,
" Dana, Logan, and many others. The traveller
" amid the Alps, and other mountain chains, will
" there see clear and unmistakable signs of such
" former catastrophes, each of which resulted from
" fractures utterly inexplicable by reference to any
" of those puny oscillations of the earth, which can
" be appealed to during historical times." . . . .
" Again,[1] I see in existing nature no cause of
" sufficient intensity to account for ordinary sedi-
" ments (once charged with organic remains) having
" been changed into crystalline masses occupying

[1] *Siluria ;* 1867 Edition, page 495.

" whole regions. The theorist in vain endeavours
" to explain such operations by processes so slow
" in their action, as to be almost imperceptible. If
" it be argued that the strata constituting lofty
" mountains were metamorphosed in parts by such
" a slow process, let any one who sustains that
" view explain how it is that every stratum in a
" lofty range of mountains, composed of carbon-
" ate of lime, should, in some cases, all at once
" change into sulphate of lime, and in others into
" dolomite."

24. Sir Charles Lyell himself admits a warmer
climate in the earliest geological periods. Thus
considering "a general[1] refrigeration of climate;"
(from the more ancient times understood) "and
" several oscillations of temperature during the
" glacial epoch;" to be proved by palæontological
evidence; he endeavours to explain those past
changes chiefly if not solely by hypothetical alterations in the distribution of land and sea over the
globe. Every reader of the *Principles of Geology*
must admire the ingenuity, and admit the import-

---

[1] *Principles of Geology*, Vol. I., page 387. 1867 Edition.

ance, of the chapter in which this hypothesis is set forth. But I earnestly beg Professor Huxley, and those in whose name he speaks, to reconsider their opinion (§ 4 above), that the secular cooling of the earth and of the sun " has made no practical " difference to the earth during the period of " which a record is preserved in stratified deposits." There is, surely, good ground for Sir Roderick Murchison's opinion that metamorphic causes have been more active in ancient times than at present, because of more rapid augmentation of temperature downwards below the earth's surface ; and it cannot be reasonably urged that a hotter sun is not a probable explanation of the supposed warmer climate of the palæozoic ages.

25. The "grave charge of opposition to the principles of Natural Philosophy," which Professor Huxley so earnestly repudiates, was carefully limited by the words in which I expressed it, to certain clearly specified points ; and it was only because of the prominent and fundamental position given to those points in many of our standard works, that I brought that charge against " British

Popular Geology." I have no wish to press the charge, merely for the sake of proving myself to have been in the right at the time I made it ; and if it rested solely on the question of geological time, I would willingly avoid repeating it. But in some of the most recent geological writings of the highest character I still find the same tendency to overlook essential principles of thermodynamics, as that to which I called the attention [1] of the geological section of the British Association, at Manchester in 1861.

26. In the last edition of *The Principles of Geology*, 1868, vol. 2, page 242, we find the following statement :—" The existence of electrical " currents in the earth's crust, and the changes in " direction which they may undergo after great " geological revolutions in the position of mountain " chains, and of land and sea, the connection also " of solar and terrestrial magnetism, and of this " last with electricity and chemical action, may

---

[1] In a communication published afterwards, under the title, "Secular Cooling of the Earth," in the *Transactions of the Royal Society of Edinburgh*, 1862, and in Thomson and Tait's *Natural Philosophy*, Appendix D. (1867).

" help us to conceive *such a cycle of change as may*
" *restore to the planet the heat supposed to be lost by*
" *radiation into space.*" And again, at page 213—
" It is a favourite dogma of some physicists, that
" not only the earth, but the sun itself, is con-
" tinually losing a portion of its heat, and that, as
" there is no known source by which it can be
" restored, we can foresee the time when all life
" will cease to exist upon this planet ; and, on
" the other hand, we can look back to the period
" when the heat was so intense as to be incom-
" patible with the existence of any organic beings
" such as are known to us in the living or fossil
" world."

" When we consider the discoveries recently
" made, of the convertibility of one kind of force
" into another, and how light, heat, magnetism,
" electricity, and chemical affinity are intimately
" connected, we may well hesitate before we accept
" this theory of the constant diminution from age
" to age of a great source of dynamical and vital
" power." These statements are directly opposed
to the general principle of the dissipation of

energy: and the hypothesis which they suggest is very inconsistent with our special knowledge of the conduction and radiation of heat, of thermo-electric currents, of chemical action, and of physical astronomy.

Kant's hypothesis of the restoration of a new chaos, like the old one, with potential energy for a repetition of cosmogony, described by Professor Huxley, was not a more violent contravention of thermodynamic law; but the synthesis of its fallaciousness is more obvious.

27. Professor Huxley's own statement as to catastrophism and uniformitarianism is open to the objection of violating the principle of the conservation of energy. " Catastrophism has insisted " upon the existence of a practically unlimited " bank of force, on which the theorist might draw ; " and it has cherished the idea of the development " of the earth from a state in which its form, and " the forces which it exerted, were very different " from those we now know." . . . .

" Uniformitarianism, on the other hand, has, " with equal justice, insisted upon a practically un-

" limited bank of time, ready to discount any
"quantity of hypothetical paper."

In the Catastrophism of Leibnitz, Newton,
Sedgwick, Phillips, Hopkins, Forbes, Murchison,
and many other true geologists, which is in no
respect different as a geological doctrine from that
now described by Professor Huxley under the new
name " evolutionism," there has been no "unlimited
bank of force."   And it is because the whole
amount of energy existing in the earth has always
been essentially finite, that physical science supports their theory, and rejects, as radically opposed
to the principles of natural philosophy,   the
uniformitarianism described by Professor Huxley
in the passage just quoted.

28. Professor Huxley concludes thus : " My
"functions, as your advocate, are at an end.   I
" speak with more than the sincerity of a mere
" advocate when I express the belief that the case
"against us has entirely broken down.   The cry
" for reform which has been raised from without,
" is superfluous, inasmuch as we have long been
" reforming from within with all needful speed ;

"and the critical examination of the grounds upon
"which the very grave charge of opposition to the
"principles of Natural Philosophy has been brought
"against us, rather shows that we have exercised
"a wise discrimination in declining to meddle with
"our foundations at the bidding of the first passer-
"by, who fancies our house is not so well built as
"it might be."

The quotations which I have given above prove
that my call for reform was very far indeed from
being superfluous, and that what Professor Huxley
describes as a "reforming from within," has been
for the last ten or fifteen years in the wrong
direction, so far as the estimation of geological
time is concerned : and they bear out my statement, that modern British popular geology, "as
"represented by a very large, very influential, and
"in many respects philosophical and sound body of
"geological investigators, constituting perhaps a
"majority of British Geologists," is, on some very
important points, in "direct opposition" to the
principles of Natural Philosophy, and of Physical
Astronomy.

29. I cannot pass from Professor Huxley's last sentence without asking, Who are the occupants of "our house," and who is the "passer-by"? Is geology not a branch of physical science? Are investigations experimental and mathematical, of underground temperature, not to be regarded as an integral part of geology? Are suggestions from astronomy and thermodynamics, when adverse to a tendency in geological speculation recently become extensively popular in England through the brilliancy and eloquence of its chief promoters, to be treated by geologists as an invitation to meddle with their foundations, which a "wise discrimination" declines? For myself, I am anxious to be regarded by geologists, not as a mere passer-by, but as one constantly interested in their grand subject, and anxious, in any way, however slight, to assist them in their search for truth.

PART II.—ON THE ORIGIN AND TOTAL AMOUNT OF PLUTONIC ENERGY.

30. By Plutonic action, I mean any disturbance of underground equilibrium. Volcanoes, earth-quakes, and subsidences are the phenomena most commonly understood when plutonic activity is spoken of. The store of energy to which these phenomena are due is properly called plutonic energy, and according to the clear and simple, but thoroughly rigorous, language of modern dynamics, plutonic energy is to be distinguished from plutonic activity.

31. The *action* of a dynamical agent was defined by Newton, as something to be measured numerically, by the number measuring simple force or pressure, multiplied into the number measuring the velocity with which the matter experiencing it yields in the direction of the force. In the nineteenth century dynamical vocabulary, Newton's "action of an agent" is simply a *performing of*

work, and we distinguish between action, or rate of action, as defined by Newton, and the integral amount of action or integral amount of work done after any operation of force is completed. Again, in modern physical dynamics we have learned that every performance of work consists in merely a transformation or intertransposition of materials, or a stopping of some motion and generating of other instead, and that when work is performed in one locality, another locality must on that account be left with so much less of the wherewithal for the farther performance of work. This "wherewithal" is called energy; and thus the performance of work is simply the drawing of energy from one store and laying it out elsewhere. Any irreversible transformation of energy is called a dissipation of energy; of which the most prominent examples are the conduction of heat from warmer to colder parts of a body, or of the matter occupying any portion of space, and the generation of heat by friction or collision.

32. Plutonic action is, therefore, to be defined as any transformation of energy going on within the

earth. No natural operation is thoroughly reversible, and therefore, every plutonic action involves
something in the way of dissipation of energy.
But the grand and awful phenomena of volcanoes
and earthquakes, results of abnormal plutonic
activity, give rise probably to much less dissipation
of energy, summed for all parts of the earth from
age to age, than the continual silent action of the
conduction of heat outwards, the amount of which
we are able to estimate in a thoroughly definite
manner. Thus we find that from year to year the
earth, at the present time, is parting with heat at
the rate of 92 horse-power [1] per square kilometre.[2]
That is to say, from a square metre of surface the
loss of energy is at an average rate of seven metre-

[1] "One horse-power" is a rate of performing work equal to
(33,000 foot pounds, or) 4·563 metre-tons per minute ; the French
ton of 1000 kilogrammes understood, being ·9842 of the British
ton.

[2] The kilometre is ·62138 of that very inconvenient measure, the
British statute mile. The square kilometre is 247·11 of that, if possible worse measure, the acre. Experts can tell how many square
yards are in an acre : but of all the men in England accustomed to
reckon their land in acres, and to state, or read, or hear reckonings
of political statistics in square miles, very few could readily answer
the question, How many acres are there in a square mile ?

tons per million seconds, or 220 metre-tons per
annum. The whole area of the earth is 510,000,000
square kilometres ; and therefore the loss from the
whole earth is 3600 millions of metre-tons per
second, or 112×10^{15} metre-tons per annum. This
statement is not hypothetical in any respect. But
the numerical data assumed in it, being 005
gramme-water-units per centimetre per second for
conductivity, and 1° cent. per 30 metres for the
rate of increase of underground temperature downwards, are what Professor Huxley would justly call
loose, because we do not know the true average
conductivity of the upper strata for the whole
earth, nor the true average value of the rate of
augmentation of temperature per metre downwards ; and a considerable margin of probable error
must be allowed for any estimate that can yet be
made of the true rate at which energy is being lost
from the earth. This, however, does not at all
affect the principles in illustration of which I
adduce the numbers, or the importance of these
principles for the success of geology as a science.

33. The store of energy, transformations of

which constitute plutonic action, consists certainly at the present time in a great measure, if not altogether, of terrestrial heat. This indeed is the only description of energy *proved* to exist in any considerable quantity within the earth; but it is possible that there may be great masses of uncombined chemical elements, and that the potential energy of their mutual affinities may constitute a considerable portion of the plutonic energy in store, whether for the generation of future underground heat, or for immediate application to some of the more violent manifestations of plutonic activity. Now, there are two ways of estimating the possible total amount of plutonic energy; one by taking the earth as it is, and not reasoning from antecedent conditions, but simply estimating from known properties of matter; how much heat it is conceivable may exist in it in its present condition; the other by tracing the history of the earth backwards.

34. From experiments such as have not yet been made, but could be made with very great ease, on the total heats of fusion of ordinary rocks and

metals,[1] we shall probably soon be able to estimate, without any very unsatisfactory degree of vagueness, a limit to the possible amount of heat in the earth. With a view to putting together the data required for this estimate, it is important to notice that we have strong reason to believe the earth is not a mere thin shell filled with melted material of rock or metal, or both, as many French and a few English geologists assume it to be; but is solid from surface to centre with the exception of comparatively small spaces still occupied by fluid lava, or subjected occasionally to melting in volcanic action.[2] We may therefore say it is not at all probable that there is now within the earth a hundred times as much heat as that which would raise a quantity of average surface rock equal in mass to the whole earth, from zero to 200° cent., since this would be certainly many times more

---

[1] A very simple plan would be to pour small quantities of melted rock into hollows in blocks of cast iron, massive enough not to rise more than a few degrees of temperature, by the communication of heat from the melted rocks.

[2] "Rigidity of the Earth" (W. Thomson), *Trans. R.S.*, 1862 ; and Thomson and Tait's *Natural Philosophy*, §§ 832-849.

than enough to melt that amount of any kind of surface rock under any moderate pressure. But merely from consideration of thermal capacities, and possible temperatures of the earth at great depths, we are not at present able to make any much less vague estimate than that, of the possible total amount of heat.

35. Inasmuch as energy is being continually lost from the earth by conduction through the upper strata, the whole quantity of plutonic energy must have been greater in past times than at present, and the question forces itself upon us, how was it first acquired? As the earth, being finite, cannot ever have had an infinite store of energy within it, there must have been a time when it was not a warm body, parting with energy, as it is now. If the matter of the earth existed before that time, it must have been under conditions which led to its being warm, and to its commencing to part with energy. It may have gained its heat by communication from other matter, or by work performed upon it by matter not now forming part of itself. But the only probable hypothesis is, that it has

become warm by the conversion of mutual potential energy, whether of gravitational, or gravitational and chemical, attraction between its parts,
into heat.

36. It may be said, why not admit previous
kinetic energy without limit, as we have no reason
to believe that the antecedent condition of the
matter now constituting the earth was a condition
of rest rather than a condition of motion? I
answer that we know nothing of absolute motion
or rest in the universe, and that any great degree
of *relative* motion of different portions of matter
through space, renders the chances of their hitting
one another very small. I therefore say it is not
probable that the portions of matter now constituting the earth had in their antecedent condition
any great amount of relative motion; and it is
probable that the kinetic energy which was converted into heat in their coalition was the equivalent
of kinetic energy acquired by mutual gravitation.
It seems, indeed, that Kant's "attempt to account
" for the constitution and mechanical origin of the
" universe, on Newtonian principles," only wanted

the knowledge of thermodynamics, which the subsequent experiments of Davy, Rumford, and Joule supplied, to lead to a thoroughly definite explanation of all that is known regarding the present actions and temperatures of the earth, and of the sun, and other heavenly bodies. And if Carnot's theory had been before him, he assuredly would not have forestalled Hutton in the chimera of "a reproductive operation, by which a ruined constitution may be repaired."[1]

37. Now the whole amount of potential energy exhausted in the coming together of the earth's materials, from infinite mutual distances (that is to say, from distances many times greater than the present diameter of the earth) to their present relative positions, is easily estimated with great accuracy with the knowledge we possess of the earth's average density. If the density were uniform from surface to centre, the amount of potential energy in question would be equal to the work

---

[1] See the account of Kant's Cosmogony given by Professor Huxley, in his "Address" of Feb. 19, 1869, to the London Geological Society.

required to lift a body equal to $\frac{3}{5}$ of the earth's mass from the present surface to an infinite distance. But observation proves the mean density of the earth to be 5·6, which is about twice the average surface density; and if we use Laplace's probable law of interior density,[1] we find more exhaustion of energy in coalition, by about 10 per cent., than if the density were uniform, the result for the whole being, as nearly as may be, a mass equal to $\frac{2}{3}$ of the earth's, raised from the surface to an infinite distance. This second estimate we may adopt with great confidence, as probably very close to the truth, considering how little it differs from the first. Now, the work required to lift a mass from the earth's surface to an infinite distance, against the diminishing force of gravity, is the same as that which would be required to lift an equal mass through a space equal to the earth's radius, against a force everywhere equal to the actual force of gravity at the surface. Hence, as the earth's radius is 6,370 kilometres, the whole amount of potential energy exhausted in the

[1] Thomson and Tait, § 824.

coalition of its parts amounts to $\frac{2}{3} \times 6,370,000$ or 4,250,000 metre-tons per ton of its whole mass: the metre-ton (an ordinary gravitation unit of work) being the amount of work required to overcome, through a space of one metre, a force equal to the weight of a ton at the earth's surface; the difference of the force of gravity at different parts of the earth's surface neglected. But unless, which is very improbable, the conglomeration took place quite suddenly by the simultaneous collision of materials falling in from all sides, a large part of this energy must have been dissipated away by radiation of heat consequent on partial collisions. We must therefore look on the definite estimate 4,250,000 metre-tons per ton of the earth's mass, which expresses somewhat accurately the whole potential energy exhausted during the conglomeration, as being considerably above the greatest amount of plutonic energy due to gravitation, that can ever have existed in the earth at any one time.

38. To estimate the potential energy of chemical affinity already exhausted, or yet to be exhausted,

by the combination of the materials constituting the earth, we may remark first, that the upper crust consists chiefly of metallic oxides, but contains also a large quantity of carbonic acid and water. Now we have the following results, from two very accurate observers, regarding heat of combination—reduced so as to show the amount of heat generated per unit mass of the compound substance formed :—

HEAT OF COMBINATION OF VARIOUS ELEMENTS WITH OXYGEN.

Substance.	Product.	Quantity of Substance.	Quantity of Oxygen.	Units of Heat Evolved.	Observer.
Potassium	K O	$\frac{39}{55}$	$\frac{16}{55}$	1682	Joule.
Iron . .	Fe$_3$ O$_4$	$\frac{21}{29}$	$\frac{8}{29}$	1141	Andrews.
Carbon .	C O^2	$\frac{3}{11}$	$\frac{8}{11}$	2155	Do.
Hydrogen	H$_2$ O	$\frac{1}{9}$	$\frac{8}{9}$	3756	Do.
Zinc . .	Zn O	$\frac{65\cdot3}{81\cdot3}$	$\frac{16}{81\cdot3}$	1045	Do.
Tin . .	Sn O$_2$	$\frac{118}{150}$	$\frac{32}{150}$	969	Do.
Copper	Cn O	$\frac{127}{159}$	$\frac{32}{159}$	481	Do.

These numbers make it, I think, very certain that the heat of combination per ton of the average materials of the earth would be over estimated at 3,000 units centigrade—that is, 3,000 times the quantity of heat required to raise the temperature of a ton of water by 1° cent., or, according to Joule's equivalent, 1,270,000 metre-tons of energy.

39. The number 4,250,000 previously found (§ 37) for the amount of potential energy of gravitation exhausted in the coalition of the earth's mass, is $3\frac{1}{3}$ times this estimate of the potential energy of the chemical affinity of its elements. The whole amount of energy due to the two causes together is about $5\frac{1}{2}$ million metre-tons, or 13,000 thermal units centigrade, per ton of the earth's mass. This, being about 700 times as much heat as would raise the temperature of an equal mass of surface rock from 0° to 100° cent., is three and a-half times the amount stated in § 34, as an over-estimate of the whole amount of heat at present in the earth. But considering, as in § 37, how much heat must have been dissipated during the conglomeration of the materials which now constitute the earth, we are rather compelled to contract than permitted to enlarge our ideas of

the possible total of plutonic energy at present in the earth, by tracing its history backwards to its probable origin.

PART III.—NOTE ON THE METEORIC THEORY
OF THE SUN'S HEAT.

[*From Report of the Glasgow Philosophical Society's
Meeting of March* 24, 1869.]

40. SIR WM. THOMSON, in reply to a question from the President, Dr. Bryce, said that his contribution to the meteoric theory of solar heat had been to point out that the meteoric supply could not be perennial. In his paper " On the Mechanical Energies of the Solar System " (*Trans. R.S.E.*, April, 1854: republished in *Math.* and *Phys. Papers*, Vol. II., pp. 1–25), he had shown that meteors falling from extra-planetary space in sufficient abundance to generate the heat emitted from the sun for the last 2000 years, must, by the augmentation they must have brought to the central mass, have caused a gradual shortening of the year of which the accumulated effect during that period must have dislocated the

seasons to the extent of a month and a half. But observation proves that there has been a dislocation of the seasons only to the extent of about an hour and three-quarters, since a certain eclipse of the moon was seen on March 19, 721 B.C., in Babylon. It is quite certain, therefore, that meteoric supply for sun heat has not within historical periods come from distant space outside the earth's orbit. He therefore found it necessary to modify the meteoric hypothesis of sun-heat—a hypothesis which he had learned from a communication by Mr. Waterston to the British Association at Hull in 1853, but which he has since found had been previously proposed by Mayer. If it is true that the heat emitted by the sun is compensated from year to year by meteors, he proved that instead of a certain quantity of meteors falling in a certain time from distant extra-planetary space, as supposed by Mayer and Waterston, a double quantity in the same time must fall from orbits inside that of Mercury. But at the same time he pointed out that observation and dynamical theory of the motions of the planets must be had recourse to, to test whether or not there can be a sufficient amount of matter circulating as meteors inside the orbit of Mercury to provide sun-heat for a few hundred years to come. Since that time Leverrier's fine researches on the motions of the planet Mercury give evidence of matter circulating as a great number of small planets within his orbit round the sun. But the amount of matter thus indicated is very small, probably not enough for a few hundred years' heat. It is therefore highly improbable that the heat of the sun depends at all for its continuation upon a continued meteoric supply. In the present state of science what appears most probable is Helmholtz's view, that the sun originally acquired his heat in being built up out of smaller masses falling together and generating heat by their collision, but that at present he is simply an incandescent mass cooling. In an article in *Macmillan's Magazine*, March, 1862, " On the Age of the Sun's Heat," he (Sir W. Thomson) had shown that the sun may have been several million future years giving out heat and light from the vast initial supply generated in that manner ; but that, without supposing the sun to be a miraculous body, continually violating the laws of matter, we cannot believe that from first to last he could illuminate the earth for several times one hundred million years, if even for so long a period as that. Since he had been asked to explain his views regarding the theory of sun-heat, he took the opportunity of adverting to a statement which Professor Huxley had recently made in his inaugural address to the Geological Society of London, to the effect that he (Sir W. Thomson) had only fifteen years ago entertained a view of the origin of the sun's heat which would have " suited Hutton perfectly," inasmuch as, according to that view, the energy radiating from year to year is supplied from year to year. But Professor Huxley had not noticed that the very limited supply which could possibly exist in store, according to that view, could not upon any estimate amount to three hundred thousand years' expenditure, at present rate even without taking into account the astronomical observations published since 1854. And, in fact, no view except Hegel's—" the motion of the heavenly

" bodies is not a being pulled this way and that,
" as is imagined (by the Newtonians); they go
" along, as the ancients said, like blessed gods,"
—could satisfy a " thorough-going Huttonian
uniformitarian," or could fulfil the conditions
imagined by Lyell as a foundation for a theory
of under-ground heat. As to the sun, we can now
go both backwards and forwards in his history,
upon the principles of Newton and Joule. A large
proportion of British popular geologists of the
present day have been longer contented than other
scientific men, to look upon the sun as Fonte-
nelle's roses looked upon their gardener.[1] " Our
" gardener," say they, " must be a very old man ;
" within the memory of roses he is the same as he
" has always been ; it is impossible he can ever die,
" or be other than he is."

[1] Kant's *Physische Geographie* (Collected Works, vol. vi.,
Leipzig, 1839).

PRESIDENTIAL ADDRESS

TO THE

BRITISH ASSOCIATION,

EDINBURGH, 1871.

FOR the third time of its forty years' history the British Association is assembled in the metropolis of Scotland. The origin of the Association is connected with Edinburgh in undying memory through the honoured names of Robison, Brewster, Forbes, and Johnston.

In this place, from this Chair, twenty-one years ago, Sir David Brewster said :—" On the return of " the British Association to the metropolis of Scot- " land I am naturally reminded of the small band " of pilgrims who carried the seeds of this Institu- ' tion into the more genial soil of our sister land. . . . . . "Sir John Robison, Professor Johnston " and Professor J. D. Forbes were the earliest

" friends and promoters of the British Association.
" They went to York to assist in its establishment,
" and they found there the very men who were
" qualified to foster and organise it. The Rev.
" Mr. Vernon Harcourt, whose name cannot be
" mentioned here without gratitude, had provided
" laws for its government, and, along with Mr.
" Phillips, the oldest and most valuable of our
" office-bearers, had made all those arrangements
" by which its success was insured. Headed by
" Sir Roderick Murchison, one of the very earliest
" and most active advocates of the Association,
" there assembled at York about two hundred of
" the friends of science."

The statement I have read contains no allusion
to the real origin of the British Association. This
blank in my predecessor's historical sketch I am
able to fill in from words written by himself
twenty years earlier. Through the kindness of
Professor Phillips I am enabled to read to you
part of a letter to him at York, written by David
Brewster from Allerly by Melrose, on the 23rd of
February, 1831 :—

" DEAR SIR ;—I have taken the liberty of writing
" you on a subject of considerable importance. It
" is proposed to establish a British Association of
" men of science similar to that which has existed
" for eight years in Germany, and which is now
" patronised by the most powerful Sovereigns of
" that part of Europe. The arrangements for the
" first meeting are in progress ; and it is contem-
" plated that it shall be held in York, as the most
" central city for the three kingdoms. My object
" in writing you at present is to beg that you
" would ascertain if York will furnish the accom-
" modation necessary for so large a meeting (which
" may perhaps consist of above one hundred
" individuals), if the Philosophical Society would
" enter zealously into the plan, and if the Mayor
" and influential persons in the town and in the
" vicinity would be likely to promote its objects.
" The principal object of the Society would be to
" make the cultivators of science acquainted with
" each other, to stimulate one another to new
" exertions, and to bring the objects of science
" more before the public eye, and to take measures

" for advancing its interests and accelerating its
" progress."

Of the little band of four pilgrims from Scotland
to York, not one now survives. Of the seven first
Associates one more has gone over to the majority
since the Association last met. Vernon Harcourt
is no longer with us; but his influence remains;
à beneficent and surely therefore never dying
influence. He was a Geologist and Chemist, a
large-hearted lover of science, and an unwearied
worker for its advancement. Brewster was the
founder of the British Association; Vernon
Harcourt was its law-giver. His code remains
to this day the law of the Association.

On the eleventh of May last Sir John Herschel
died in the eightieth year of his age. The name
of Herschel is a household word throughout Great
Britain and Ireland—yes, and through the whole
civilised world. We of this generation have, from
our lessons of childhood upwards, learned to see in
Herschel, father and son, a *præsidium et dulce decus*
of the precious treasure of British scientific fame.
When geography, astronomy, and the use of the

globes were still taught, even to poor children, as a pleasant and profitable sequel to "reading, writing, and arithmetic," which of us did not revere the great telescope of Sir William Herschel (one of the Hundred Wonders of the World), and learn with delight, directly or indirectly from the charming pages of Sir John Herschel's book, about the sun and his spots, and the fiery tornadoes sweeping over his surface, and about the planets, and Jupiter's belts, and Saturn's rings, and the fixed stars with their proper motions, and the double stars, and coloured stars, and the nebulæ discovered by the great telescope? Of Sir John Herschel it may indeed be said, *nil tetigit quod non ornavit.*

With regard to Sir John Herschel's scientific work, on the present occasion I can but refer briefly to a few points which seem to me salient in his physical and mathematical writings. First, I remark that he has put forward, most instructively and profitably to his readers, the general theory of periodicity in dynamics, and has urged the practical utilising of it, especially in meteorology, by the harmonic analysis. It is purely by an application of this principle and practical method, that the British Association's Committee on Tides has for the last four years been, and still is, working towards the solution of the grand problem proposed forty-eight years ago by Thomas Young in the following words ;—

"There is, indeed, little doubt that if we were "provided with a sufficiently correct series of "minutely accurate observations on the Tides, "made not merely with a view to the times of low "and high water only, but rather to the heights "at the intermediate times, we might form, by "degrees, with the assistance of the theory con-"tained in this article[1] only, almost as perfect a "set of tables for the motions of the ocean as we "have already obtained for those of the celestial "bodies, which are the more immediate objects of "the attention of the practical astronomer."

Sir John Herschel's discovery of a right or left-handed asymmetry in the outward form of crystals such as quartz, which in their inner molecular

[1] Young's ; written in 1823 for the Supplement to the *Encyclopædia Britannica.*

structure possess the heliçoidal rotational property in reference to the plane of polarisation of light is one of the notable points of meeting between Natural History and Natural Philosophy. His observations on "epipolic dispersion" gave Stokes the clue by which he was led to his great discovery of the change of periodic time experienced by light in falling on certain substances and being dispersively reflected from them. In respect to pure mathematics Sir John Herschel did more, I believe, than any other man to introduce into Britain the powerful methods and the valuable notation of modern analysis. A remarkable mode of symbolism had freshly appeared, I believe, in the works of Laplace, and possibly of other French mathematicians; it certainly appeared in Fourier but whether before or after Herschel's work I cannot say. With the French writers, however, this was rather a short method of writing formulæ than the analytical engine which it became in the hands of Herschel and British followers, especially Sylvester and Gregory (competitors with Green in the Cambridge Mathematical Tripos struggle of

1837) and Boole and Cayley. This method was greatly advanced by Gregory, who first gave to its working-power a secure and philosophical foundation and so prepared the way for the marvellous extension it has received from Boole, Sylvester and Cayley, according to which symbols of operation become the subjects not merely of algebraic combination, but of differentiations and integrations as if they were symbols expressing values of varying quantities.   An even more marvellous development of this same idea of the separation of symbols (according to which Gregory separated the algebraic signs + and − from other symbols or quantities to be characterised by them, and dealt with them according to the laws of algebraic combination) received from Hamilton a most astonishing generalisation, by the invention actually of new laws of combination, and led him to his famous "Quaternions," of which he gave his earliest exposition to the Mathematical and Physical Section of this Association, at its meeting in Cambridge in the year 1845.   Tait has taken up the subject of quaternions ably and zealously, and has carried it into physical science with a faith shared by some of the most thoughtful mathematical naturalists of the day, that it is destined to become an engine of perhaps hitherto unimagined power for investigating and expressing results in Natural Philosophy. Of Herschel's gigantic work in astronomical observation I need say nothing. Doubtless a careful account of it will be given in the *Proceedings of the Royal Society of London* for the next anniversary meeting.

In the past year another representative man of British science is gone. Mathematics has had no steadier supporter for half a century than De Morgan. His great book on the differential calculus was, for the mathematical student of thirty years ago, a highly prized repository of all the best things that could be brought together under that title. I do not believe it is less valuable now; and if it is less valued, may this not be because it is too good for examination purposes, and because the modern student, labouring to win marks in the struggle for existence, must not suffer himself to be beguiled from the stern path of duty by any attractive beauties in the subject of his study?

One of the most valuable services to science which the British Association has performed has been the establishment, and the twenty-nine years' maintenance, of its Observatory. The Royal Meteorological Observatory of Kew was built originally for a Sovereign of England who was a zealous amateur of astronomy. George the Third used continually to repair to it when any celestial phenomenon of peculiar interest was to be seen ; and a manuscript book still exists filled with observations written into it by his own hand. After the building had been many years unused, it was granted, in the year 1842, by the Commissioners of Her Majesty's Woods and Forests, on application of Sir Edward Sabine, for the purpose of continuing observations (from which he had already deduced important results) regarding the vibration of a pendulum in various gases, and for the purpose of promoting pendulum observations in all parts of the world. The Government granted only the building—no funds for carrying on the work to be done in it. The Royal Society was unable to undertake the maintenance of such an observatory ; but, happily for science, the zeal of individual Fellows of the Royal Society and Members of the British Association gave the initial impulse, supplied the necessary initial funds, and recommended their new institution successfully to the fostering care of the British Association. The work of the Kew Observatory has, from the commencement, been conducted under the direction of a Committee of the British Association ; and annual grants from the funds of the Association have been made towards defraying its expenses up to the present time. To the initial object of pendulum research was added continuous observation of the phenomena of meteorology and terrestrial magnetism, and the construction and verification of thermometers, barometers, and magnetometers designed for accurate measurement. The magnificent services which it has rendered to science are so well known that any statement of them which I could attempt on the present occasion would be superfluous. Their value is due in a great measure to the indefatigable zeal and the great ability of two Scotchmen, both from Edinburgh, who successively held the office of Superintendent of the Observatory of the British

Association—Mr. Welsh for nine years, until his death in 1859, and Dr. Balfour Stewart from then until the present time.   Fruits of their labours are to be found all through our volumes of Reports for these twenty-one years.

The institution now enters on a new stage of its existence.   The noble liberality of a private benefactor, one who has laboured for its welfare with self-sacrificing devotion unintermittingly from within a few years of its creation, has given it a permanent independence, under the general management of a Committee of the Royal Society. Mr. Gassiot's gift of £10,000 secures the continuance at Kew of the regular operation of the self-recording instruments for observing the phenomena of terrestrial magnetism and meteorology, without the necessity for further support from the British Association.

The success of the Kew Magnetic and Meteorological Observatory affords an example of the great gain to be earned for science by the foundation of physical observatories and laboratories for experimental research, to be conducted by qualified persons, whose duties should be, not teaching, but experimenting. Whether we look to the honour of England, as a nation which ought always to be the foremost in promoting physical science, or to those vast economical advantages which must accrue from such establishments, we cannot but feel that experimental research ought to be made with us an object of national concern, and not left, as hitherto, exclusively to the private enterprise of self-sacrificing amateurs, and the necessarily inconsecutive action of our present Governmental Departments and of casual Committees. The Council of the Royal Society of Edinburgh has moved for this object in a memorial presented by them to the Royal Commission on Scientific Education and the Advancement of Science. The Continent of Europe is referred to for an example to be followed with advantage in this country, in the following words :—

" On the Continent there exist certain institu-" tions, fitted with instruments, apparatus, ' chemicals, and other appliances, which are meant " to be, and which are made, available to men of " science, to enable them, at a moderate cost, to " pursue original researches."

This statement is fully corroborated by information, on good authority, which I have received
from Germany, to the effect that in Prussia " every
" university, every polytechnical academy, every in-
" dustrial school (Realschule and Gewerbeschule),
" most of the grammar-schools, in a word, nearly
" all the schools superior in rank to the elementary
" schools of the common people, are supplied with
" chemical laboratories and a collection of philo-
" sophical instruments and apparatus, access to
" which is most liberally granted by the directors of
" those schools, or the teachers of the respective
" disciplines, to any person qualified, for *scientific*
" *experiments.*   In consequence, though there exist
" no particular institutions like those mentioned
" in the memorial there will scarcely be found a
" town exceeding in number 5,000 inhabitants but
" offers the possibility of *scientific explorations* at
" no other cost than reimbursement of the expense
" for the materials wasted in the experiments."

Further, with reference to a remark in the Memorial to the effect that, in respect to the pro-
motion of science, the British Government confines

its action almost exclusively to scientific instruction, and fatally neglects the advancement of science, my informant tells me that, in Germany, " professors, preceptors, and teachers of secondary " schools are engaged on account of their skilfulness " in *teaching;* but professors of universities are " never engaged unless they have already proved, " *by their own investigations,* that they are to be " relied upon for the *advancement of science.* There-" fore every shilling spent for instruction in "universities is at the same time profitable to the " advancement of science."

The physical laboratories which have grown up in the Universities of Glasgow and Edinburgh, and in Owens College, Manchester, show the want felt of Colleges of Research ; but they go but infinitesimally towards supplying it, being absolutely destitute of means, material or personal, for advancing science except at the expense of volunteers, or for securing that volunteers shall be found to continue even such little work as at present is carried on.

The whole of Andrews' splendid work in Queen's

College, Belfast, has been done under great difficulties and disadvantages, and at great personal
sacrifices ; and up to the present time there is not
a student's physical laboratory in any one of the
Queen's Colleges in Ireland—a want which surely
ought not to remain unsupplied.   Each of these
institutions (the four Scotch Universities, the three
Queen's Colleges, and Owens College, Manchester)
requires two professors of Natural Philosophy—one
who shall be responsible for the teaching, the other
for the advancement of science by experiment.
The University of Oxford has already established
a physical laboratory.   The munificence of its
Chancellor is about to supply the University of
Cambridge with a splendid laboratory, to be
constructed under the eye of Professor Clerk
Maxwell.   On this subject I shall say no more at
present, but simply read a sentence which was
spoken by Lord Milton in the first Presidential
Address to the British Association, when it met at
York in the year 1831 :—" In addition to other
" more direct benefits, these meetings [of the
" British Association], I hope, will be the means

" of impressing on the Government the conviction
" that the love of scientific pursuits, and the means
" of pursuing them, are not confined to the metro-
" polis ; and I hope that when the Government is
" fully impressed with the knowledge of the great
" desire entertained to promote science in every
" part of the empire, they will see the necessity of
" affording it due encouragement, and of giving
" every proper stimulus to its advancement."

Besides abstracts of papers read, and discussions
held, before the Sections, the annual Reports of the
British Association contain a large mass of valuable
matter of another class. It was an early practice
of the Association, a practice that might well be
further developed, to call occasionally for a special
report on some particular branch of science from a
man eminently qualified for the task. The reports
received in compliance with these invitations have
all done good service in their time, and they remain
permanently useful as landmarks in the history of
science. Some of them have led to vast practical
results ; others of a more abstract character are
valuable to this day as powerful and instructive

condensations and expositions of the branches of science to which they relate. I cannot better illustrate the two kinds of efficiency realised in this department of the Association's work than by referring to Cayley's "Report on Abstract Dynamics"[1] and Sabine's "Report on Terrestrial Magnetism"[2] (1838).

To the great value of the former, personal experience of benefit received enables me, and gratitude impels me, to testify. In a few pages full of precious matter, the generalised dynamical equations of Lagrange, the great principle evolved from Maupertuis' "least action" by Hamilton, and the later developments and applications of the Hamiltonian principle by other authors are described by Cayley so suggestively that the reading of thousands of quarto pages of papers scattered through the *Transactions* of the various learned Societies of Europe is rendered superfluous for

[1] "Report on the Recent Progress of Theoretical Dynamics," by A. Cayley (*Report of the British Association*, 1857, p. 1).

[2] "Report on the Variations of the Magnetic Intensity observed at different points of the Earth's Surface," by Major Sabine, F.R.S. (forming part of the seventh *Report of the British Association*).

any one who desires only the essence of these investigations, with no more of detail than is necessary for a thorough and practical understanding of the subject.

Sabine's Report of 1838 concludes with the following sentence:—" Viewed in itself and its " various relations, the magnetism of the earth " cannot be counted less than one of the most " important branches of the physical history of the " planet we inhabit ; and we may feel quite assured " that the completion of our knowledge of its " distribution on the surface of the earth would be " regarded by our contemporaries and by posterity " as a fitting enterprise of a maritime people, and " a worthy achievement of a nation which has ever " sought to rank foremost in every arduous and " honourable undertaking." An immediate result of this Report was that the enterprise which it proposed was recommended to the Government by a joint Committee of the British Association and the Royal Society with such success, that Captain James Ross was sent in command of the *Erebus* and *Terror* to make a magnetic survey of the

Antarctic regions, and to plant on his way three Magnetical and Meteorological Observatories, at St. Helena, the Cape, and Van Diemen's Land. A vast mass of precious observations, made chiefly on board ship, were brought home from this expedition. To deduce the desired results from them, it was necessary to eliminate the disturbance produced by the ship's magnetism; and Sabine asked his friend Archibald Smith to work out from Poisson's mathematical theory, then the only available guide, the formulæ required for the purpose. This voluntary task Smith executed skilfully and successfully. It was the beginning of a series of labours carried on with most remarkable practical tact, with thorough analytical skill, and with a rare extreme of disinterestedness, in the intervals of an arduous profession, for the purpose of perfecting and simplifying the correction of the mariner's compass—a problem which had become one of vital importance for navigation, on account of the introduction of iron ships. Edition after edition of the *Admiralty Compass Manual* has been produced by the able superintendent of the Compass

Department, Captain Evans, containing chapters of mathematical investigation and formulæ by Smith, on which depend wholly the practical analysis of compass-observations, and rules for the safe use of the compass in navigation. I firmly believe that it is to the thoroughly scientific method thus adopted by the Admiralty, that no iron ship of Her Majesty's Navy has ever been lost through errors of the compass. The *British Admiralty Compass Manual* is adopted as a guide by all the navies of the world. It has been translated into Russian, German, and Portuguese; and it is at present being translated into French. The British Association may be gratified to know that the possibility of navigating ironclad war-ships with safety depends on application of scientific principles given to the world by three mathematicians, Poisson, Airy, and Archibald Smith.

Returning to the science of terrestrial magnetism we find in the Reports of early years of the British Association ample evidence of its diligent cultivation. Many of the chief scientific men of the day from England, Scotland, and Ireland found a

strong attraction to the Association in the facilities
which it afforded to them for co-operating in their
work on this subject. Lloyd, Phillips, Fox, Ross,
and Sabine made magnetic observations all over
Great Britain ; and their results, collected by
Sabine, gave for the first time an accurate and
complete survey of terrestrial magnetism over the
area of this island. I am informed by Professor
Phillips that, in the beginning of the Association,
Herschel, though a " sincere well-wisher," felt
doubts as to the general utility and probable
success of the plan and purpose proposed ; but his
zeal for terrestrial magnetism brought him from
being merely a sincere well-wisher to join actively
and cordially in the work of the Association. " In
" 1838 he began to give effectual aid in the great
" question of magnetical Observatories, and was
" indeed foremost among the supporters of that
" which is really Sabine's great work. At intervals,
" until about 1858, Herschel continued to give
" effectual aid." Sabine has carried on his great
work without intermission to the present day ;
thirty years ago he gave to Gauss a large part of

the data required for working out the spherical harmonic analysis of terrestrial magnetism over the whole earth. A recalculation of the harmonic analysis for the altered state of terrestrial magnetism of the present time has been undertaken by Adams. He writes to me that he has " already " begun some of the introductory work, so as to be " ready when Sir Edward Sabine's Tables of the " values of the Magnetic Elements deduced from " observation are completed, at once to make use " of them," and that he intends to take into account terms of at least one order beyond those included by Gauss. The form in which the requisite data are to be presented to him is a magnetic Chart of the whole surface of the globe. Materials from scientific travellers of all nations, from our home magnetic observatories, from the magnetic observatories of St. Helena, the Cape, Van Diemen's Land, and Toronto, and from the scientific observatories of other countries have been brought together by Sabine. Silently, day after day, night after night, for a quarter of a century he has toiled with one constant assistant always by his side to reduce these observations and prepare for the great work. At this moment, while we are here assembled, I believe that, in their quiet summer retirement in Wales, Sir Edward and Lady Sabine are at work on the magnetic Chart of the world. If two years of life and health are granted to them, science will be provided with a key which must powerfully conduce to the ultimate opening up of one of the most refractory enigmas of cosmical physics, the cause of terrestrial magnetism.

To give any sketch, however slight, of scientific investigation performed during the past year would, even if I were competent for the task, far exceed the limits within which I am confined on the present occasion. A detailed account of work done and knowledge gained in science Britain ought to have every year. The *Journal of the Chemical Society* and the *Zoological Record* do excellent service by giving abstracts of all papers published in their departments. The admirable example afforded by the German *Fortschritte* and *Jahresbericht* is before us ; but hitherto, so far as I know, no attempt has been made to follow it in Britain. It is true that several of the annual volumes of the *Jahresbericht* were translated ; but a translation, published necessarily at a considerable interval of time after the original, cannot supply the want. An independent British publication is for many obvious reasons desirable. The two publications, in German and English, would, both by their differences and by their agreements, illustrate the progress of science more correctly and usefully than any single work could do, even if appearing simultaneously in the two languages. It seems to me that to promote the establishment of a British Year Book of Science is an object to which the powerful action of the British Association would be thoroughly appropriate.

In referring to recent advances in several branches of science, I simply choose some of those which have struck me as most notable.

Accurate and minute measurement seems to the non-scientific imagination a less lofty and dignified work than looking for something new. But nearly all the grandest discoveries of science have been but the rewards of accurate measurement and patient long-continued labour in the minute sifting of numerical results. The popular idea of Newton's grandest discovery is that the theory of gravitation flashed into his mind, and so the discovery was made. It was by a long train of mathematical calculation, founded on results accumulated through prodigious toil of practical astronomers, that Newton first demonstrated the forces urging the planets towards the Sun, determined the magnitudes of those forces, and discovered that a force following the same law of variation with distance urges the Moon towards the Earth. *Then* first, we may suppose, came to him the idea of the universality of gravitation ; but when he attempted to compare the magnitude of the force on the Moon with the magnitude of the force of gravitation of a heavy body of equal mass at the earth's surface, he did not find the agreement which the law he was discovering required. Not for years after would he publish his discovery as made. It is recounted that, being present at a meeting of the Royal Society, he heard a paper read, describing geodesic measurement by Picard which led to a serious correction of the previously accepted estimate of the Earth's radius. This was what Newton required. He went home with the result, and commenced his calculations, but felt so much agitated that he handed over the arithmetical work to a friend : then (and not when, sitting in a garden, he saw an apple fall) did he ascertain that gravitation keeps the Moon in her orbit.

Faraday's discovery of specific inductive capacity, which inaugurated the new philosophy, tending to discard action at a distance was the result of minute and accurate measurement of electric forces.

Joule's discovery of thermo-dynamic law through the regions of electro-chemistry, electro-magnetism, and elasticity of gases was based on a delicacy of thermometry which seems simply impossible to some of the most distinguished chemists of the day.

Andrews' discovery of the continuity between the gaseous and liquid states was worked out by many years of laborious and minute measurement of phenomena scarcely sensible to the naked eye.

Great service has been done to science by the British Association in promoting accurate measurement in various subjects. The origin of exact science in terrestrial magnetism is traceable to Gauss's invention of methods of finding the magnetic intensity in absolute measure. I have spoken of the great work done by the British Association in carrying out the application of this invention in all parts of the world. Gauss's colleague in the German Magnetic Union, Weber, extended the practice of absolute measurement to electric currents, the resistance of an electric conductor, and the electromotive force of a galvanic element. He showed the relation between electrostatic and electromagnetic units for absolute measurement, and made the beautiful discovery that resistance, in absolute electromagnetic measure, and the reciprocal of resistance, or, as we call it "conducting power," in electrostatic measure, are each of them a velocity. He made an elaborate and difficult series of experiments to measure the velocity which is equal to the conducting power, in electrostatic measure, and at the same time to the resistance in electromagnetic measure, in one and the same conductor. Maxwell, in making the first advance along a road of which Faraday was the pioneer, discovered that this velocity is physically related to the velocity of light, and that, on a certain hypothesis regarding the elastic medium concerned, it may be exactly equal to the velocity of light. Weber's measurement verifies approximately this equality, and stands in science *monumentum ære perennius*, celebrated as having suggested this most grand theory, and as having afforded the first quantitative test of the recondite properties of matter on which the relations between electricity and light depend. A remeasurement of Weber's critical velocity on a new plan by Maxwell himself, and the important correction of the velocity of light by Foucault's laboratory experiments, verified by astronomical observation, seem to show a still closer agreement. The most accurate possible determination of Weber's critical velocity is just now a primary object of the Association's Committee on Electric Measurement; and it is at present premature to speculate as to the closeness of the agreement between that velocity and the velocity of light. This leads me to remark how much science, even in its most lofty speculations, gains in return for benefits conferred by its application to promote the social and material welfare of man. Those who perilled and lost their money in the original Atlantic Telegraph were impelled and supported by a sense of the grandeur of their enterprise, and of the world-wide benefits which must flow from its success ; they were at the same time not unmoved by the beauty of the scientific problem directly presented to them ; but they little thought that it was to be immediately, through their work, that the scientific world was to be instructed in a long-neglected and discredited fundamental electric discovery of Faraday's, or that, again, when the assistance of the British Association was invoked to supply their electricians with methods for absolute measurement (which they found necessary to secure the best economical return for their expenditure, and to obviate and detect those faults in their electric material which had led to disaster), they were laying the foundation for accurate electric measurement in every scientific laboratory in the world, and initiating a train of investigation which now sends up branches into the loftiest regions and subtlest ether of natural philosophy. Long may the British Association continue a bond of union, and a medium for the interchange of good offices between science and the world !

The greatest achievement yet made in molecular theory of the properties of matter is the Kinetic theory of Gases, shadowed forth by Lucretius, definitely stated by Daniel Bernoulli, largely developed by Herapath, made a reality by Joule, and worked out to its present advanced state by Clausius and Maxwell. Joule, from his dynamical equivalent of heat, and his experiments upon the heat produced by the condensation of gas, was able to estimate the average velocity of the ultimate molecules or atoms composing it. His estimate for hydrogen was 6225 feet per second at temperature 60° Fahr., and 6055 feet per second at the freezing-point. Clausius took fully into account the impacts of molecules on one another, and the kinetic energy of *relative* motions of the matter constituting an individual atom. He investigated the relation between their diameters, the number in a given space, and the mean length of path from impact to impact, and so gave the foundation for estimates of the absolute dimensions of atoms, to which I shall refer later. He explained the slowness of gaseous diffusion by the mutual impacts of the atoms, and laid a secure foundation for a complete theory of the diffusion of fluids, previously a most refractory enigma. The deeply penetrating genius of Maxwell brought in viscosity and thermal conductivity, and thus completed the dynamical explanation of all the known properties of gases, except their electric resistance and brittleness to electric force.

No such comprehensive molecular theory had ever been even imagined before the nineteenth century. Definite and complete in its area as it is, it is but a well-drawn part of a great chart, in which all physical science will be represented with every property of matter shown in dynamical relation to the whole. The prospect we now have of an early completion of this chart is based on the assumption of atoms. But there can be no permanent satisfaction to the mind in explaining heat, light, elasticity, diffusion, electricity and magnetism, in gases, liquids, and solids, and describing precisely the relations of these different states of matter to one another by statistics of great numbers of atoms, when the properties of the atom itself are simply assumed. When the theory, of which we have the first instalment in Clausius and Maxwell's work, is complete, we are but brought face to face with a superlatively grand question, What is the inner mechanism of the atom?

In the answer to this question we must find the explanation not only of the atomic elasticity, by which the atom is a chronometric vibrator according to Stokes's discovery, but of chemical affinity and of the differences of quality of different chemical elements, at present a mere mystery in science. Helmholtz's exquisite theory of vortex-motion in an incompressible frictionless liquid has been suggested as a finger-post, pointing a way which may possibly lead to a full understanding

of the properties of atoms, carrying out the grand
conception of Lucretius, who "admits no subtle
" ethers, no variety of elements with fiery, or
" watery, or light, or heavy principles ; nor supposes
" light to be one thing, fire another, electricity a
" fluid, magnetism a vital principle, but treats all
" phenomena as mere properties or accidents of
" simple matter." This statement I take from an
admirable paper [by Fleeming Jenkin] on the
atomic theory of Lucretius, which appeared in
the *North British Review* for March 1868, containing a most interesting and instructive sum-
mary of ancient and modern doctrine regarding
atoms. Allow me to read from that article
one other short passage finely describing the
present aspect of atomic theory :—" The exist-
" ence of the chemical atom, already quite a
" complex little world, seems very probable ;
" and the description of the Lucretian atom is
" wonderfully applicable to it. We are not wholly
" without hope that the real weight of each such
" atom may some day be known—not merely the
" relative weight of the several atoms, but the
" number in a given volume of any material ; that

" the form and motion of the parts of each atom
" and the distances by which they are separated
" may be calculated; that the motions by which
" they produce heat, electricity, and light may be
" illustrated by exact geometrical diagrams; and
" that the fundamental properties of the interme-
" diate and possibly constituent medium may be
" arrived at. Then the motion of planets and
" music of the spheres will be neglected for a
" while in admiration of the maze in which the
" tiny atoms run."

Even before this was written some of the
anticipated results had been partially attained.
Loschmidt in Vienna had shown, and not
much later Stoney independently in England
showed, how to deduce from Clausius and Maxwell's kinetic theory of gases a superior limit
to the number of atoms in a given measurable
space. I was quite unaware of what Loschmidt
and Stoney had done when I made a similar
estimate on the same foundation, and communicated it to *Nature* in an article on "The Size
of Atoms." But questions of personal priority,
however interesting they may be to the persons

concerned, sink into insignificance in the prospect
of any gain of deeper insight into the secrets of
nature.   The triple coincidence of independent
reasoning in this case is valuable as confirmation
of a conclusion violently contravening ideas and
opinions which had been almost universally held
regarding the dimensions of the molecular structure
of matter.   Chemists and other naturalists had
been in the habit of evading questions as to the
hardness or indivisibility of atoms by virtually
assuming them to be infinitely small and infinitely
numerous.   We must now no longer look upon the
atom, with Boscovich, as a mystic point endowed
with inertia and the attribute of attracting or
repelling other such centres with forces depending
upon the intervening distances (a supposition only
tolerated with the tacit assumption that the inertia
and attraction of each atom is infinitely small and
the number of atoms infinitely great), nor can we
agree with those who have attributed to the atom
occupation of space with infinite hardness and
strength (incredible in any finite body); but we
must realise it as a piece of matter of measurable

dimensions, with shape, motion, and laws of action, intelligible subjects of scientific investigation.

The prismatic analysis of light discovered by Newton was estimated by himself as "the odd- " est, if not the most considerable, detection " which hath hitherto been made in the operations " of nature." Had he not been deflected from the subject, [had he had nineteenth century optical glass for his prisms] he could not have failed to obtain a pure spectrum ; but this, with the inevitably consequent discovery of the dark lines, was reserved for the nineteenth century. Our fundamental knowledge of the dark lines is due to Fraunhofer. Wollaston saw them, but did not discover them. Brewster laboured long and well to perfect the prismatic analysis of sunlight ; and his observations on the dark bands produced by the absorption of interposed gases and vapours laid foundations for the grand superstructure which he scarcely lived to see. Piazzi Smyth, by spectroscopic observation performed on the Peak of Teneriffe, added greatly to our knowledge of the dark lines produced in the solar spectrum by the absorption of our own atmosphere. The prism became an instrument for chemical qualitative analysis in the hands of Fox Talbot and Herschel, who first showed how, through it, the old "blowpipe test" or generally the estimation of substances from the colours which they give to flames, can be prosecuted with an accuracy and a discriminating power not to be attained when the colour is judged by the unaided eye. But the application of this test to solar and stellar chemistry had never, I believe, been suggested, either directly or indirectly, by any other naturalist, when Stokes taught it to me in Cambridge at some time prior to the summer of 1852. The observational and experimental foundations on which he built were :—

(1) The discovery by Fraunhofer of a coincidence between his double dark line D of the solar spectrum and a double bright line which he observed in the spectra of ordinary artificial flames.

(2) A very rigorous experimental test of this coincidence by Prof. W. H. Miller, which showed

it to be accurate to an astonishing degree of minuteness.

(3) The fact that the yellow light given out when salt is thrown on burning spirit consists almost solely of the two nearly identical qualities which constitute that double bright line.

(4) Observations made by Stokes himself, which showed the bright line D to be absent in a candle-flame when the wick was snuffed clean, so as not to project into the luminous envelope, and from an alcohol flame when the spirit was burned in a watch-glass. And

(5) Foucault's admirable discovery (*L'Institut,* Feb. 7, 1849) that the voltaic arc between charcoal points is " a medium which emits the rays D on its " own account, and at the same time absorbs them " when they come from another quarter."

The conclusions, theoretical and practical, which Stokes taught me, and which I gave regularly afterwards in my public lectures in the University of Glasgow, were :—

(1) That the double line D, whether bright or dark, is due to vapour of sodium.

(2) That the ultimate atom of sodium is sus-
ceptible of regular elastic vibrations, like those of a
tuning-fork or of stringed musical instruments ;
that like an instrument with two strings tuned to
approximate unison, or an approximately circular
elastic disk, it has two fundamental notes or vibrations of approximately equal pitch ; and that the
periods of these vibrations are precisely the periods
of the two slightly different yellow lights constituting the double bright line D.

(3) That when vapour of sodium is at a high
enough temperature to become itself a source of
light, each atom executes these two fundamental
vibrations simultaneously ; and that therefore the
light proceeding from it is of the two qualities
constituting the double bright line D.

(4) That when vapour of sodium is present in
space across which light from another source is
propagated, its atoms, according to a well-known
general principle of dynamics, are set to vibrate
in either or both of those fundamental modes, if
some of the incident light is of one or other of
their periods, or some of one and some of the

other ; so that the energy of the waves of those particular qualities of light is converted into thermal vibrations of the medium and dispersed in all directions, while light of all other qualities, even though very nearly agreeing with them, is transmitted with comparatively no loss.

(5) That Fraunhofer's double dark line D of solar and stellar spectra is due to the presence of vapour of sodium in atmospheres surrounding the sun and those stars in whose spectra it had been observed.

(6) That other vapours than sodium are to be found in the atmospheres of sun and stars by searching for substances producing in the spectra of artificial flames bright lines coinciding with other dark lines of the solar and stellar spectra than the Fraunhofer line D.

The last of these propositions I felt to be confirmed (it was perhaps partly suggested) by a striking and beautiful experiment admirably adapted for lecture illustrations, due to Foucault, which had been shown to me by M. Duboscque Soleil, and the Abbé Moigno, in Paris in the month of October 1850. A prism and lenses were arranged to throw upon a screen an approximately pure spectrum of a vertical electric arc between charcoal poles of a powerful battery, the lower one of which was hollowed like a cup. When pieces of copper and pieces of zinc were separately thrown into the cup, the spectrum exhibited, in perfectly definite positions, magnificent well-marked bands of different colours characteristic of the two metals. When a piece of brass, compounded of copper and zinc, was put into the cup, the spectrum showed all the bands, each precisely in the place in which it had been seen when one metal or the other had been used separately.

It is much to be regretted that this great generalisation was not published to the world twenty years ago. I say this, not because it is to be regretted that Ångström should have the credit of having in 1853 published independently the statement that " an incandescent gas emits lumi- " nous rays of the same refrangibility as those " which it can absorb " ; or that Balfour Stewart should have been unassisted by it when, coming to the subject from a very different point of view, he made, in his extension of the " Theory of Exchanges,"[1] the still wider generalisation that the radiating power of every kind of substance is equal to its absorbing power for every kind of ray ; or that Kirchhoff also should have in 1859 independently discovered the same proposition, and shown its application to solar and stellar chemistry ; but because we might now be in possession of the inconceivable riches of astronomical results which we expect from the next ten years' investigation by spectrum analysis, had Stokes given his theory to the world when it first occurred to him.

To Kirchhoff belongs, I believe, solely the great credit of having first actually sought for and found other metals than sodium in the sun by the method of spectrum analysis. His publication of October 1859 inaugurated the practice of solar and stellar chemistry, and gave spectrum analysis an impulse to which in a great measure is due its splendidly successful cultivation by the labours of many able investigators within the last ten years.

[1] *Edin. Transactions*, 1858-59.

To the prodigious and wearing toil of Kirchhoff himself, and of Angström, we owe large-scale maps of the solar spectrum, incomparably superior in minuteness and accuracy of delineation to any thing ever attempted previously. These maps now constitute the standards of reference for all workers in the field. Plücker and Hittorf opened ground in advancing the physics of spectrum analysis and made the important discovery of changes in the spectra of ignited gases produced by changes in the physical condition of the gas. The scientific value of the meetings of the British Association is well illustrated by the fact that it was through conversation with Plücker at the Newcastle meeting that Lockyer was first led into the investigation of the effects of varied pressure on the quality of the light emitted by glowing gas which he and Frankland have prosecuted with such admirable success. Scientific wealth tends to accumulation according to the law of compound interest. Every addition to knowledge of properties of matter supplies the naturalist with new instrumental means for discovering and interpreting phenomena of nature, which in their turn afford foundations for fresh generalisations, bringing gains of permanent value into the great storehouse of philosophy. Thus Frankland, led, from observing the want of brightness of a candle burning in a tent on the summit of Mont Blanc, to scrutinise Davy's theory of flame, discovered that brightness without incandescent solid particles is given to a purely gaseous flame by augmented pressure, and that a dense ignited gas gives a spectrum comparable with that of the light from an incandescent solid or liquid. Lockyer joined him ; and the two found that every incandescent substance gives a continuous spectrum—that an incandescent gas under varied pressure gives bright bars across the continuous spectrum, some of which, from the sharp, hard and fast lines observed where the gas is in a state of extreme attenuation, broaden out on each side into nebulous bands as the density is increased, and are ultimately lost in the continuous spectrum when the condensation is pushed on till the gas becomes a fluid no longer to be called gaseous. More recently they have examined the influence of temperature, and have obtained results which seem to show that a highly attenuated gas, which at a high temperature gives several bright lines, gives a smaller and smaller number of lines, of sufficient brightness to be visible, when the temperature is lowered, the density being kept unchanged. I cannot refrain here from remarking how admirably this beautiful investigation harmonises with Andrews' great discovery of continuity between the gaseous and liquid states. Such things make the life-blood of science. In contemplating them we feel as if led out from narrow waters of scholastic dogma to a refreshing excursion on the broad and deep ocean of truth, where we learn from the wonders we see that there are endlessly more and more glorious wonders still unseen.

Stokes' dynamical theory supplies the key to the philosophy of Frankland and Lockyer's discovery. Any atom of gas when struck and left to itself vibrates with perfect purity its fundamental note or notes. In a highly attenuated gas each atom is very rarely in collision with other atoms, and therefore is nearly at all times in a state of true

vibration. Hence the spectrum of a highly
attenuated gas consists of one or more perfectly
sharp bright lines, with a scarcely perceptible
continuous gradation of prismatic colour. In
denser gas each atom is frequently in collision,
but still is for much more time free, in intervals between collisions, than engaged in collision ;
so that not only is the atom itself thrown sensibly
out of tune during a sensible proportion of its
whole time, but the confused jangle of vibrations in
every variety of period during the actual collision
becomes more considerable in its influence. Hence
bright lines in the spectrum broaden out somewhat,
and the continuous spectrum becomes less faint.
In still denser gas each atom may be almost as
much time in collision as free, and the spectrum
then consists of broad nebulous bands crossing a
continuous spectrum of considerable brightness.
When the medium is so dense that each atom is
always in collision, that is to say never free from
the influence of its neighbours, the spectrum will
generally be continuous, and may present little or
no appearance of bands, or even of maxima of

brightness. In this condition the fluid can be no longer regarded as a gas, and we must judge of its relation to the vaporous or liquid states according to the critical conditions discovered by Andrews.

While these great investigations of properties of matter were going on, naturalists were not idle with the newly recognised power of the spectroscope at their service. Chemists soon followed the example of Bunsen in discovering new metals in terrestrial matter by the old blow-pipe and prism test of Fox Talbot and Herschel. Biologists applied spectrum analysis to animal and vegetable chemistry, and to sanitary investigations. But it is in astronomy that spectroscopic research has been carried on with the greatest activity, and been most richly rewarded with results. The chemist and the astronomer have joined their forces. An astronomical observatory has now, appended to it, a stock of reagents such as hitherto was only to be found in the chemical laboratory. A devoted corps of volunteers of all nations, whose motto might well be *ubique*, have directed their artillery to every region of the universe. The sun, the

N 2

spots on his surface, the corona and the red and yellow prominences seen round him during total eclipses, the moon, the planets, comets, auroras, nebulæ, white stars, yellow stars, red stars, variable and temporary stars, each tested by the prism was compelled to show its distinguishing colours. Rarely before in the history of science has enthusiastic perseverance directed by penetrative genius produced within ten years so brilliant a succession of discoveries. It is not merely the *chemistry* of sun and stars, as first suggested, that is subjected to analysis by the spectroscope. Their whole laws of being are now subjects of direct investigation ; and already we have glimpses of their evolutional history through the stupendous power of this most subtle and delicate test. We had only solar and stellar chemistry ; we now have solar and stellar physiology.

It is an old idea that the colour of a star may be influenced by its motion relatively to the eye of the spectator, so as to be tinged with red if it moves from the earth, or blue if it moves towards the earth. William Allen Miller, Huggins, and

Maxwell showed how, by aid of the spectroscope, this idea may be made the foundation of a method of measuring the relative velocity with which a star approaches to or recedes from the earth. The principle is, first to identify, if possible, one or more of the lines in the spectrum of the star, with a line or lines in the spectrum of sodium, or some other terrestrial substance, and then (by observing the star and the artificial light simultaneously by the same spectroscope) to find the difference, if any, between their refrangibilities. From this difference of refrangibility the ratio of the periods of the two lights is calculated, according to data determined by Fraunhofer from comparisons between the positions of the dark lines in the prismatic spectrum and in his own "interference spectrum" (produced by substituting for the prism a fine grating). A first comparatively rough application of the test by Miller and Huggins to a large number of the principal stars of our skies, including Aldebaran, a Orionis, β Pegasi, Sirius, a Lyræ, Capella, Arcturus, Pollux, Castor (which they had observed rather for the chemical purpose than for this), proved that not one of them had so great a velocity as 315 kilometres per second to or from the earth, which is a *most momentous result in respect to cosmical dynamics.* Afterwards Huggins made special observations of the velocity test, and succeeded in making the measurement in one case, that of Sirius, which he then found to be receding from the earth at the rate of 66 kilometres per second. This, corrected for the velocity of the earth at the time of the observation, gave a velocity of Sirius, relatively to the Sun, amounting to 47 kilometres per second. The minuteness of the difference to be measured, and the smallness of the amount of light, even when the brightest star is observed, renders the observation extremely difficult. Still, with such great skill as Mr. Huggins has brought to bear on the investigation, it can scarcely be doubted that velocities of many other stars may be measured. What is now wanted is, certainly not greater skill, perhaps not even more powerful instruments, but *more instruments and more observers.* Lockyer's applications of the velocity test to the relative motions of different gases in the Sun's photosphere, spots, chromosphere, and chromospheric prominences, and his observations of the varying spectra presented by the same substance as it moves from one position to another in the Sun's atmosphere and his interpretations of these observations, according to the laboratory results of Frankland and himself, go far towards confirming the conviction that in a few years all the marvels of the Sun will be dynamically explained according to known properties of matter.

During six or eight precious minutes of time, spectroscopes have been applied to the solar atmosphere and to the corona seen round the dark disk of the Moon eclipsing the Sun. Some of the wonderful results of such observations, made in India on the occasion of the eclipse of August 1868, were described by Professor Stokes in a previous address. Valuable results have, through the liberal assistance given by the British and American Governments, been obtained also from the total eclipse of last December, notwithstanding a generally unfavourable condition of weather. It seems to have been proved that at least some sensible part of the light of the "corona" is a terrestrial atmospheric halo or dispersive reflection of the light of the glowing hydrogen and "helium" [1] round the sun. I believe I may say, on the present occasion, when preparation must again be made to utilise a total eclipse of the Sun, that the British Association confidently trusts to our Government exercising the same wise liberality as heretofore in the interests of science.

The old nebular hypothesis supposes the solar system and other similar systems through the universe which we see at a distance as stars, to have originated in the condensation of fiery nebulous matter. This hypothesis was invented before the discovery of thermodynamics, or the nebulæ would not have been supposed to be fiery ; and the idea seems never to have occurred to any of its inventors or early supporters that the matter, the condensation of which they supposed to consti-

---

[1] Frankland and Lockyer find the yellow prominences to give a very decided bright line not far from D, but hitherto not identified with any terrestrial flame. It seems to indicate a new substance, which they propose to call Helium.

tute the Sun and stars, could have been other than fiery in the beginning. Mayer first suggested that the heat of the Sun may be due to gravitation : but he supposed meteors falling in to keep always generating the heat which is radiated year by year from the Sun. Helmholtz, on the other hand, adopting the nebular hypothesis, showed in 1854 that it was not necessary to suppose the nebulous matter to have been originally fiery, but that mutual gravitation between its parts may have generated the heat to which the present high temperature of the Sun is due. Further, he made the important observations that the potential energy of gravitation in the Sun is even now far from exhausted ; but that with further and further shrinking more and more heat is to be generated, and that thus we can conceive the Sun even now to possess a sufficient store of energy to produce heat and light, almost as at present, for several million years of time future. It ought, however, to be added that this condensation can only follow from cooling, and therefore that Helmholtz's gravitational explanation of future Sun-heat amounts really to showing that the Sun's thermal capacity is enormously greater, in virtue of the mutual gravitation between the parts of so enormous a mass, than the sum of the thermal capacities of separate and smaller bodies of the same material and same total mass. Reasons for adopting this theory, and the consequences which follow from it, are discussed in an article " On the Age of the Sun's Heat," published in *Macmillan's Magazine* for March 1862.

For a few years Mayer's theory of solar heat had seemed to me probable ; but I had been led to regard it as no longer tenable, because I had been in the first place driven, by consideration of the very approximate constancy of the Earth's period of revolution round the Sun for the last 2000 years, to conclude that " The principal source, " perhaps the sole appreciably effective source of " Sun-heat, is in bodies circulating round the Sun at present inside the Earth's orbit"; [1] and because

[1] " On the Mechanical Energies of the Solar System," *Transactions of the Royal Society of Edinburgh*, 1854 ; and *Phil. Mag.* 1854, second half year.

Leverrier's researches on the motion of the planet Mercury, though giving evidence of a sensible influence attributable to matter circulating as a great number of small planets within his orbit round the Sun, showed that the amount of matter that could possibly be assumed to circulate at any considerable distance from the Sun must be very small ; and therefore, " if the meteoric influx " taking place at present is enough to produce any " appreciable portion of the heat radiated away, it " must be supposed to be from matter circulating " round the Sun, within very short distances of " his surface. The density of this meteoric cloud " would have to be supposed so great that comets " could scarcely have escaped, as comets actu- " ally have escaped, showing no discoverable " effects of resistance, after passing his sur- " face within a distance equal to one-eighth of " his radius. All things considered, there seems " little probability in the hypothesis that solar " radiation is compensated to any appreciable " degree by heat generated by meteors falling in, " at present ; and, as it can be shown that no

" chemical theory is tenable,[1] it must be concluded
" as most probable that the Sun is at present
" merely an incandescent liquid mass cooling." [2]

Thus on purely astronomical grounds was I long
ago led to abandon as very improbable the
hypothesis that the Sun's heat is supplied dynamically from year to year by the influx of meteors.
But now spectrum analysis gives proof finally
conclusive against it.

Each meteor circulating round the Sun must
fall in along a very gradual spiral path, and before
reaching the Sun must have been for a long time
exposed to an enormous heating effect from his
radiation when very near, and must thus have been
driven into vapour before actually falling into the
Sun. Thus, if Mayer's hypothesis is correct friction between vortices of meteoric vapours and the
Sun's atmosphere must be the immediate cause of
solar heat ; and the velocity with which these
vapours circulate round equatorial parts of the
Sun must amount to 435 kilometres per second.

---

[1] " Mechanical Energies," &c.
[2] " Age of the Sun's Heat " (*Macmillan's Magazine*, March, 1862).

The spectrum test of velocity applied by Lockyer showed but a twentieth part of this amount as the greatest observed relative velocity between different vapours in the Sun's atmosphere.

At the first Liverpool Meeting of the British Association (1854), in advancing a gravitational theory to account for all the heat, light, and motions of the universe, I urged that the immediately antecedent condition of the matter of which the Sun and planets were formed, not being fiery, could not have been gaseous ; but that it probably was solid, and may have been like the meteoric stones which we still so frequently meet with through space.   The discovery of Huggins, that the light of the Nebulæ, so far as hitherto sensible to us, proceeds from incandescent hydrogen and nitrogen gases, and that the heads of comets also give us light of incandescent gas, seems at first sight literally to fulfil that part of the Nebular hypothesis to which I had objected.   But a solution, which seems to me in the highest degree probable, has been suggested by Tait.   He supposes that it may be by ignited gaseous exhalations proceeding from the collision of meteoric stones that Nebulæ, and the heads of Comets, show themselves to us, and he suggested, at a former meeting of the Association, that experiments should be made for the purpose of applying spectrum analysis to the light which has been observed in gunnery trials, such as those at Shoeburyness, when iron strikes against iron at a great velocity, but varied by substituting for the iron various solid materials, metallic or stony. Hitherto this suggestion has not been acted upon ; but surely it is one the carrying out of which ought to be promoted by the British Association.

Most important steps have been recently made towards the discovery of the nature of comets ; establishing with nothing short of certainty the truth of a hypothesis which had long appeared to me probable,—that they consist of groups of meteoric stones ;—accounting satisfactorily for the light of the nucleus ; and giving a simple and rational explanation of phenomena presented by the tails of comets which had been regarded by the greatest astronomers as almost preternaturally marvellous. The meteoric hypothesis to which I have referred remained a mere hypothesis (I do not know that it was ever even published), until, in 1866, Schiaparelli calculated, from observations on the August meteors, an orbit for these bodies which he found to agree almost perfectly with the orbit of the great comet of 1862 as calculated by Oppolzer ; and so discovered and demonstrated that a comet consists of a group of meteoric stones. Professor Newton, of Yale College, United States, by examining ancient records, ascertained that in periods of about thirty-three years, since the year 902, there have been exceptionally brilliant displays of the November meteors. It had long been believed that these interesting visitants came from a train of small detached planets circulating round the Sun all in nearly the same orbit, and constituting a belt analogous to Saturn's ring, and that the reason for the comparatively large number of meteors which we observe annually about the 14th of November is, that at that time the earth's orbit cuts through the supposed meteoric belt. Professor Newton concluded from his investigation that there is a denser part of the group of meteors which extends over a portion of the orbit so great as to occupy about one-tenth or one-fifteenth of the periodic time in passing any particular point, and gave a choice of five different periods for the revolution of this meteoric stream round the sun, any one of which would satisfy his statistical result. He further concluded that the line of nodes, that is to say, the line in which the plane of the meteoric belt cuts the plane of the Earth's orbit, has a progressive sidereal motion of about $52''4$ per annum Here, then, was a splendid problem for the physical astronomer; and, happily, one well qualified for the task, took it up. Adams, by the application of a beautiful method invented by Gauss, found that of the five periods allowed by Newton just one permitted the motion of the line of nodes to be explained by the disturbing influence of Jupiter, Saturn, and other planets. The period chosen on these grounds is $33\frac{1}{4}$ years. The investigation showed further that the form of the orbit is a long ellipse, giving for shortest distance from the Sun 145 million kilometres, and for longest distance 2895 .million kilometres. Adams also worked out the longitude of the perihelion and

the inclination of the orbit's plane to the plane of
the ecliptic.   The orbit which he thus found agreed
so closely with that of Tempel's Comet I. 1866
that he was able to identify the comet and the
meteoric belt   The same conclusion had been

[1] Signor Schiaparelli, Director of the Observatory of Milan, who,
in a letter dated 31st December, 1866, pointed out that the elements
of the orbit of the *August* meteors, calculated from the observed position of their radiant point on the supposition of the
orbit being a very elongated ellipse agreed very closely with those of
the orbit of Comet II., 1862, calculated by Dr. Oppolzer.   In the
same letter Schiaparelli gives elements of the orbit of the November
meteors, but these were not sufficiently accurate to enable him to
identify the orbit with that of any known comet.   On the 21st
January, 1867, M. Leverrier gave more accurate elements of the
orbit of the November meteors, and in the *Astronomische Nach-*
richten of January 9, Mr. C. F. W. Peters, of Altona, pointed out
that these elements closely agreed with those of Tempel's Comet (I.
1866), calculated by Dr. Oppolzer, and on February 2, Schiaparelli
having recalculated the elements of the orbit of the meteors, himself
noticed the same agreement.   Adams arrived quite independently
at the conclusion that the orbit of $33\frac{1}{4}$ years period, is the one
which *must* be chosen, out of the five indicated by Prof. Newton.
His calculations were sufficiently advanced before the letters referred
to appeared, to show that the other four orbits offered by Newton
were inadmissible.   But the calculations to be gone through to
find the secular motion of the node in such an elongated orbit as that
of the meteors, were necessarily very long, so that they were not
completed till about March 1867.   They were communicated in
that month to the Cambridge Philosophical Society, and in the
month following to the Astronomical Society.

pointed out a few weeks earlier by Schiaparelli,
from calculations by himself on data supplied by
direct observations on the meteors, and independently by Peters from calculations by Leverrier on
the same foundation. It is therefore thoroughly
established that Tempel's Comet I. 1866 consists
of an elliptic train of minute planets, of which a
few thousands or millions fall to the earth annually
about the 14th of November, when we cross their
track. We have probably not yet passed through
the very nucleus or densest part; but thirteen
·times, in Octobers and Novembers, from October
13, A.D. 902 to November 14, 1866 inclusive (this
last time having been correctly predicted by Prof.
Newton), we have passed through a part of the
belt greatly denser than the average. The densest
part of the train, when near enough to us, is visible
as the head of the comet. This astounding result,
taken along with Huggins's spectroscopic observations on the light of the heads and tails of comets,
confirm most strikingly Tait's theory of comets, to
which I have already referred ; according to which
the comet, a group of meteoric stones, is self-

luminous in its nucleus, on account of collisions among its constituents, while its "tail" is merely a portion of the less dense part of the train illuminated by sunlight, and visible or invisible to us according to circumstances, not only of density, degree of illumination, and nearness, but also of tactic arrangement, as of a flock of birds or the edge of a cloud of tobacco smoke! What prodigious difficulties are to be explained, you may judge from two or three sentences which I shall read from Herschel's *Astronomy*, and from the fact that even Schiaparelli seems still to believe in the repulsion. "There is, beyond question, " some profound secret and mystery of nature " concerned in the phenomenon of their tails. " Perhaps it is not too much to hope that future " observation, borrowing every aid from rational " speculation, grounded on the progress of physical " science generally (especially those branches of " it which relate to the ethereal or imponderable " elements), may enable us ere long to penetrate " this mystery, and to declare whether it is really " *matter* in the ordinary acceptation of the term

" which is projected from their heads with such
" extraordinary velocity, and if not *impelled*, at
" least *directed*, in its course, by reference to the
" Sun, as its point of avoidance." [1]

" In no respect is the question as to the
" materiality of the tail more forcibly pressed on
" us for consideration than in that of the enormous
" sweep which it makes round the sun *in perihelio*
" in the manner of a straight and rigid rod, *in*
" *defiance of the law of gravitation*, nay, even *of*
" *the received* laws of motion." [1]

" The projection of this ray . . . to so enormous
" a length, in a single day conveys an impression
" of the intensity of the forces acting to produce
" such a velocity of material transfer through space,
" such as no other natural phenomenon is capable
" of exciting. It is clear that *if we have to deal*
" *here with matter, such as we conceive it*, viz.,
" *possessing inertia—at all*, it must be under the
" dominion of forces incomparably more energetic
" than gravitation, and quite of a different nature." [2]

[1] Herschel's *Astronomy*, § 599.
[2] Herschel's *Astronomy*, 10th Edition, § 589.

Think now of the admirable simplicity with which Tait's beautiful " sea-bird analogy," as it has been called, can explain all [?] these phenomena.

The essence of science, as is well illustrated by astronomy and cosmical physics, consists in inferring antecedent conditions, and anticipating future evolutions, from phenomena which have actually come under observation. In biology the difficulties of successfully acting up to this ideal are prodigious. The earnest naturalists of the present day are, however, not appalled or paralysed by them, and are struggling boldly and laboriously to pass out of the mere " Natural History stage " of their study, and bring zoology within the range of Natural Philosophy. A very ancient specula-tion, still clung to by many naturalists (so much so that I have a choice of modern terms to quote in expressing it) supposes that, under meteorological conditions very different from the present, dead matter may have run together or crystallised or fermented into " germs of life," or " organic cells," or " protoplasm." But science brings a vast mass of inductive evidence against this hypothesis of spontaneous generation, as you have heard from my predecessor in the Presidential chair. Careful enough scrutiny has, in every case up to the present day, discovered life as antecedent to life. Dead matter cannot become living without coming under the influence of matter previously alive. This seems to me as sure a teaching of science as the law of gravitation. I utterly repudiate, as opposed to all philosophical uniformitarianism, the assumption of "different meteorological conditions"—that is to say, somewhat different vicissitudes of temperature, pressure, moisture, gaseous atmosphere—to produce or to permit that to take place by force or motion of dead matter alone, which is a direct contravention of what seems to us biological law. I am prepared for the answer, "Our code of biological law is an expression of our "ignorance as well as of our knowledge." And I say yes: search for spontaneous generation out of inorganic materials; let any one not satisfied with the purely negative testimony of which we have now so much against it, throw himself into the inquiry. Such investigations as those of

Pasteur, Pouchet, and Bastian are among the most
interesting and momentous in the whole range of
Natural History, and their results, whether positive
or negative, must richly reward the most careful
and laborious experimenting.   I confess to being
deeply impressed by the evidence put before us by
Professor Huxley, and I am ready to adopt, as an
article of scientific faith, true through all space and
through all time, that life proceeds from life, and
from nothing but life.

How, then, did life originate on the Earth?
Tracing the physical history of the Earth backwards, on strict dynamical principles, we are
brought to a red-hot melted globe on which no life
could exist.   Hence when the Earth was first fit
for life, there was no living thing on it.   There
were rocks solid and disintegrated, water, air all
round, warmed and illuminated by a brilliant Sun,
ready to become a garden.   Did grass and trees
and flowers spring into existence, in all the fulness
of ripe beauty, by a fiat of Creative Power? or did
vegetation, growing up from seed sown, spread and
multiply over the whole Earth?   Science is bound

by the everlasting law of honour, to face fearlessly every problem which can fairly be presented to it· If a probable solution, consistent with the ordinary course of nature, can be found, we must not invoke an abnormal act of Creative Power. When a lava stream flows down the sides of Vesuvius or Etna it quickly cools and becomes solid ; and after a few weeks or years it teems with vegetable and animal life ; which, for it, originated by the transport of seed and ova and by the migration of individual living creatures. When a volcanic island springs up from the sea, and after a few years is found clothed with vegetation, we do not hesitate to assume that seed has been wafted to it through the air, or floated to it on rafts. Is it not possible, and if possible, is it not probable, that the beginning of vegetable life on the Earth is to be similarly explained ? Every year thousands, probably millions, of fragments of solid matter fall upon the Earth— whence came these fragments ? What is the previous history of any one of them ? Was it created in the beginning of time an amorphous mass ? This idea is so unacceptable that, tacitly or explicitly, all men discard it. It is often assumed that all, and it is certain that some, meteoric stones are fragments which had been broken off from greater masses and launched free into space. It is as sure that collisions must occur between great masses moving through space as it is that ships, steered without intelligence directed to prevent collision, could not cross and recross the Atlantic for thousands of years with immunity from collisions. When two great masses come into collision in space it is certain that a large part of each is melted ; but it seems also quite certain that in many cases a large quantity of débris must be shot forth in all directions, much of which may have experienced no greater violence than in-dividual pieces of rock experience in a land-slip or in blasting by gunpowder. Should the time when this Earth comes into collision with another body, comparable in dimensions to itself, be when it is still clothed as at present with vegeta-tion, many great and small fragments carrying seed and living plants and animals would un-doubtedly be scattered through space. Hence and because we all confidently believe that there are at present, and have been from time immemorial, many worlds of life besides our own, we must regard it as probable in the highest degree that there are countless seed-bearing meteoric stones moving about through space. If at the present instant no life existed upon this Earth, one such stone falling upon it might, by what we blindly call *natural* causes, lead to its becoming covered with vegetation. I am fully conscious of the many scientific objections which may be urged against this hypothesis, but I believe them to be all answerable. I have already taxed your patience too severely to allow me to think of discussing any of them on the present occasion. The hypothesis that [some] life [has actually] originated on this Earth through moss-grown fragments from the ruins of another world may seem wild and visionary; all I maintain is that it is not unscientific, [and cannot rightly be said to be improbable.]

From the Earth stocked with such vegetation as it could receive meteorically, to the Earth teeming with all the endless variety of plants and animals

which now inhabit it, the step is prodigious ; yet,
according to the doctrine of continuity, most ably
laid before the Association by a predecessor in
this Chair (Mr. Grove), all creatures now living on
earth have proceeded by orderly evolution from
some such origin.   Darwin concludes his great
work on " The Origin of Species " with the following words :—" It is interesting to contemplate an
" entangled bank clothed with many plants of
" many kinds, with birds singing on the bushes,
" with various insects flitting about, and with
" worms crawling through the damp earth, and to
" reflect that these elaborately constructed forms,
" so different from each other, and dependent on
" each other in so complex a manner, have all
" been produced by laws acting around us." . . . .
" There is grandeur in this view of life with its
" several powers, having been originally breathed
" by the Creator into a few forms or into one ; and
" that, whilst this planet has gone cycling on
" according to the fixed law of gravity, from so
  simple a beginning endless forms, most beautiful
" and most wonderful, have been and are being

"evolved." With the feeling expressed in these two sentences I most cordially sympathise. I have omitted two sentences which come between them, describing briefly the hypothesis of "the "origin of species by natural selection," because I have always felt that this hypothesis does not contain the true theory of evolution, if evolution there has been, in biology. Sir John Herschel, in expressing a favourable judgment on the hypo thesis of zoological evolution, with, however, some reservation in respect to the origin of man, objected to the doctrine of natural selection, that it was too like the Laputan method of making books, and that it did not sufficiently take into account a continually guiding and controlling intelligence. This seems to me a most valuable and instructive criticism. I feel profoundly convinced that the argument of design has been greatly too much lost sight of in recent zoological speculations. Reaction against frivolities of teleology, such as are to be found, not rarely, in the notes of learned Commentators on Paley's "Natural Theology," has I believe had a temporary effect in turning attention from the solid and irrefragable argument so well put forward in that excellent old book. But overpoweringly strong proofs of intelligent and benevolent design lie all round us, and if ever perplexities, whether metaphysical or scientific, turn us away from them for a time, they come back upon us with irresistible force, showing to us through nature the influence of a free will, and teaching us that all living beings depend on one ever-acting Creator and Ruler.

PRESIDENTIAL ADDRESS
TO THE SOCIETY OF TELEGRAPH
ENGINEERS. 1874.

[The Twentieth Ordinary General Meeting was held on Wednesday, the 14th January, 1874, Sir William Thomson, F.R.S., LL.D., President, in the Chair.

The President read his Inaugural Address as follows :—

GENTLEMEN,—I thank you most cordially for the great honour you have done me in electing me to be your President for the year 1874. Our first two Presidents, Mr. Siemens and Mr. Scudamore, in their interesting and valuable addresses, have explained the object of the Society of Telegraph Engineers, and have amply demonstrated its reason for existence. The success which it has already achieved, exceeding the most sanguine expectations of its well-wishers, must be very gratifying to its public-spirited founders, as a fruit

earned by the toil and trouble they have volun-
tarily bestowed upon it. In numbers, in popularity,
in usefulness, the Society of Telegraph Engineers
has indeed grown with telegraphic speed.

When first addressed from the presidential chair,
not quite two years ago, the Society consisted of
110 members. Since that time it has augmented
to 500: including our Postmaster-General ; the
Directors-General of the great Telegraphic Ad-
ministrations of Great Britain and India ; many of
the officers and operators of those systems and of
the great Submarine Telegraph Companies ; many
scientific men interested in the subject, although
not holding official positions in connection with
practical telegraphy ; and a list of distinguished
names constituting our honorary and foreign
members.

In his inaugural address our first president said,
" Let us hope that our joint efforts may lead us
in the direction of true scientific and practical
advancement ;" and we all know how strenuously
and effectively he has himself laboured to promote
the harmony of theory and practice, not only in

the department to which this Society is devoted, but in all branches of the grand profession of engineering, of which he is so distinguished an ornament.

Before we commence the business of the session upon which we are now entering, may I be permitted to offer a few remarks on the relations between science and practice in engineering in general, but more particularly in telegraphic engineering. Engineering may be defined as the application of practical science to man's material circumstances and means of action. As usual in classification, the nomenclature of branches of engineering is full of what the logician calls cross-divisions. Thus we have civil and military engineering, and again, civil and mechanical engineering; then architecture and building, engineering and contracting. We have, it is true, in the distinction between military and civil engineering a good logical division. Every subject of civil engineering is included in military engineering, because an army has all the wants of any large body of civilians. But military engineering includes more, because there is no civil purpose which requires rifled cannon, shot and shell, congreve rockets, hand grenades, torpedoes, ironclads, armed fortifications, mining under fire, or under liability to hand-to-hand encounter with an enemy, and field telegraphs. I have enumerated all the subjects which I can think of that belong exclusively to military engineering, and, except these, all subjects of general engineering are embraced in civil engineering, properly so called.

The division between military and civil engineering is, therefore, not properly founded on a distinction in respect of the subject-matter, but it is a true logical division in respect to the province of application. Now remark the division between civil and mechanical engineering—a distinction habitually used, as if the engineering of merchant steamers, of cotton mills, of sugar machinery, of calico printing, of letter-press printing, were not truly parts of civil engineering. I make no complaint of the ordinary language which designates as civil engineering only that which is neither military, nor concerned with mechanism otherwise than in designing and testing it, and which calls mechanical engineering the construction, daily use, and maintenance of machines. I make no complaint of the ordinary language which so designates civil engineering, and distinguishes it from mechanical engineering. I only say that it is not logical. Take, again, architecture.

Architecture is not commonly called a branch of engineering at all. I think it unfortunate that the public do not regard architecture as a branch of engineering. When architects come to regard themselves as engineers, and when the public come to expect them to act as engineers, let us hope they will give us buildings not less beautiful and not less interestingly connected with monuments and traditions of beauty from bygone ages than they give us now. But assuredly there will then be less typhoid fever. Then invalids too ill to walk, or ride, or drive out of doors, or to be benefited by the beautiful scenery of Mentone, or Corsica, or Madeira, will not be expatriated merely to avoid the evil effects of the indoor atmosphere

of England. Then people in good health will not
be stupefied by a few hours of an evening at home
in gaslight, or of a social reunion, or by one hour
of a crowded popular lecture or meeting of a
learned society. Then in our hotels, and dwelling-
houses, and clubs, we shall escape the negatively
refreshing influence of the all-pervading daily
aërial telegraph, which prematurely transmits
intelligence of distant and future dinners. The
problem of giving us within doors any prescribed
degree of temperature, with air as fresh and pure
as the atmosphere outside the house can supply,
may be not an easy problem ; but it is certainly
a problem to be solved when architecture becomes
a branch of scientific engineering.

Now as to the relations between theory and
practice in telegraphic engineering, I feel that I
have more to say respecting the reflected benefits
which electrical science gains from its practical
applications in the electric telegraph than of the
value of theory in directing, and aiding, and inter
esting the operators in every department of the
work of the electric telegraph.   In no other branch

of engineering, indeed, is high science more in-
telligently appreciated and ably applied than in
the manufacture and the use of telegraphic lines,
whether over land or under sea ; and it would be
quite superfluous for me to speak on that subject
to those whom I see before me.

But I do not know whether so much is thought
of what the electric telegraph and its workers have
already done, and may be expected yet to do, for
science in general, and particularly electricity and
magnetism. Time does not allow me to enlarge
as I would like to do on this subject. I will
merely remind those who are present of the great
advance that has been made in accurate measurement within the last fifteen years. I need not
tell you that a large part of the benefit thus
achieved for science is due to the requirements of
the practical telegraphist. Men of abstract science
were satisfied to know that absolute measurement was possible, and that a definition of
magnetic force, a definition of electric resistance,
a definition of electromotive force, and so on
through the list of numerical quantities in elec-

tricity, could each of them be stated in absolute measure.

We owe to Gauss and Weber the first great practical realization in abstract science of a system of absolute measurement ; but their principles did not extend rapidly even in the domains of abstract science where their theory was well understood, because the urgent need for its practical application was not felt. When accurate measurement in any definite unit first became prevalent was when it was required by the electric telegraph. The pioneers of science in electric telegraphy, many of whom, happily for us, still work for science and for the electric telegraph, laid down—among various perfectly definite subjects for measurement —a unit of electric resistance—that most primary one of the different things to be measured respecting electricity. I need not remind any of you of the history of electric units of resistance, or of the labours of the Committee of the British Association to bring that system of measurement into harmony with the theoretical definitions of Gauss and Weber. The benefits conferred by introducing a system of definite measurement into the working of the electric telegraph are due not solely—perhaps not even in chief—to the application of Gauss's system, but to the introduction of very accurate and definite standards of resistance and means of reproducing those standards should the originals be lost. The benefit of putting the practical standards into relation with the science of Gauss and Weber has been set forth in the successive reports of the Committee of the British Association on electric measurement, and is well known, I believe, to most of the members of the Society of Telegraph Engineers.

But what I wish to say now is that theoretical science has gained great reflected benefit from the introduction of accurate measurement of resistance into practical telegraphy.

For many years measurements were performed in the office of the telegraphic factory, and at the station-house of the telegraphic wire, the means of doing which,—perhaps I might even say the principles on which those measurements were conducted,—being still unknown throughout the scientific laboratories of Europe. The professors of science who threw out the general principle have gained a rich harvest for the seed which they sowed. They have now got back from the practical telegrapher accurate standards of measurement, and ready means of transmitting those standards and of preserving them for years and years without change, which have proved of the most extreme value to the work of the scientific laboratory. I might make similar remarks regarding electric instruments. The theory of electric instruments has been taught by those who have laboured in theoretical science ; but the zeal and ability with which the makers and users of instruments in the service of the electric telegraph have taken up the hints of science have given back to the scientific laboratory instruments of incalculable value.

But I wish rather to confine myself to looking forward to the benefits which science may derive from its practical applications in telegraph engineering, and to point out that this Society is designed by its founders to be a channel through which these benefits may flow back to science, and, on the other hand, to supply the counter-channels by which pure science may exercise its perennially beneficial influence on practice.

Time would fail me to give any such statement as would include a large part of the subject upon which I have touched ; I shall therefore confine myself strictly to one point, and that is the science of terrestrial electricity I have advisedly, not thoughtlessly, used the expression " terrestrial electricity." It is not an expression we are accustomed to. We are accustomed to " terrestrial magnetism ;" we are accustomed to " atmospheric electricity." The electric telegraph forces us to combine our ideas with reference to terrestrial magnetism and atmospheric electricity. We must look upon the earth and the air as a whole—a globe of earth and air—and consider its electricity, whether at rest or in motion. Then, as to terrestrial magnetism,—of what its relation may be to perceptible electric manifestations we at present know nothing. You all know that the earth acts as a great magnet Dr. Gilbert, of Colchester made that clear nearly 300 years ago ; but how the earth acts as a great magnet—how it is a magnet,—whether an electro-magnet in virtue of currents circulating round under the upper surface, or whether it is a magnet like a mass of steel or loadstone, we do not know.

When the phenomena of terrestrial magnetism were first somewhat accurately observed about 300 years ago, the needle pointed here in England a little to the east of north ; a few years later it pointed due north ; then, until about the year 1820, it went to the west of north ; and now it is coming back towards the north. The dip has experienced corresponding variations. The dip was first discovered by the instrument-maker Robert Norman :—an illustration, I may mention in passing, of the benefits which abstract science derives from practical applications—one of the most important fundamental discoveries of magnetism brought back to theory by the instrument-maker who made mariners' compasses. Robert Norman, in balancing his compass-cards, noticed that after they were magnetized one end dipped, and he examined the phenomenon and supported a needle about the centre of gravity, magnetised it, and discovered the dip. When the dip was first so discovered by Robert Norman it was greater than it is now. The dip has gone on decreasing, and is still decreasing. In these great changes of terrestrial magnetism, or in the mere existence of terrestrial magnetism, we have always before us one of the greatest mysteries of science—a mystery which I might almost say is to myself a subject of daily contemplation. What can be the cause of this magnetism in the interior of the earth? Rigid magnetization, like that of steel or the loadstone, has no quality in itself in virtue of which we can conceive it to migrate round in the magnetised bar. Electric currents afford the more favoured hypothesis; they are more mobile. If we can conceive electric currents at all, we may conceive them flitting about. But what sustains the electric currents? People sometimes say, heedlessly or ignorantly, that thermo-electricity does it. We have none of the elements of the problem of thermo-electricity in the state of underground temperature which could possibly explain, in accordance with any knowledge we have of thermo-electricity, how there could so be sustained currents round the earth. And if there were currents round the earth, regulated by some cause so as to give them a definite direction at one time, we are as far as ever from explaining how the channel of these currents could experience that great revolutionary variation which we know it does experience. Thus we have merely a mystery. It would be rash to suggest even an explanation. I may say that one explanation has been suggested. It was suggested by the great astronomer, Halley, that there is a nucleus in the interior of the earth, and that the mystery is explained simply by a magnet not rigidly connected with the upper crust of the earth, but revolving round an axis differing from the axis of rotation of the outer crust, and exhibiting a gradual precessional motion independent of the precessional motion of the outer rigid crust. I merely say that has been suggested. I do not ask you to judge of the probability: I would not ask myself to judge of the probability of it.

But now, I say, we look with hopefulness to the practical telegraphist for data towards a solution of this grand problem. The terrestrial magnetism is subject, as a whole, to the grand secular variation which I have indicated. But, besides that, there are annual variations and diurnal variations. Every day the needle varies from a few minutes on one side to a few minutes on the other side of its mean position, and at times there are much greater variations. What are called "magnetic storms" are of not very unfrequent occurrence. In a magnetic storm the needle will often fly twenty minutes, thirty minutes, a degree, or even as much as two or three degrees sometimes, from its proper position—if I may use that term—its proper position for the time ; that is, the position which it might be expected to have at the time according to the statistics of previous observations. I speak of the needle in general. The ordinary observation of the horizontal needle shows these phenomena. So does observation on the dip of the needle. So does observation on the total intensity of the terrestrial magnetic force. The three elements, deflection, dip, and total intensity, all vary every day with the ordinary diurnal variation, and irregularly with the magnetic storm. The magnetic storm is always associated with a visible phenomenon, which we call, habitually, electrical : aurora borealis, and, no doubt, also the aurora of the southern polar regions.

We have the strongest possible reasons for believing that aurora consists of electric currents, like the electric phenomena presented by currents of electricity through what are called vacuum tubes, through the space occupied by vacuums of different qualities in the well-known vacuum tubes. Of course, the very expression "vacuums of different qualities" is a contradiction in terms. It implies that there are small quantities of matter of different kinds left in those nearest approaches to a perfect vacuum which we can make.

It is known to you all that aurora borealis is properly comparable with the phenomena presented by vacuum tubes. The appearance of the light the variations which it presents, and the magnetic accompaniments, are all confirmatory of this view so that we may accept it as one of the truths of science. Well now—and here is a point upon which, I think, the practical telegraphist not only can, but will, before long, give to abstract science data for judging—is the deflection of the needle a direct effect of the auroral current, or are the auroral current and the deflection of the needle common results of another cause ? With reference to this point, I must speak of underground currents. There, again, I have named a household word to every one who has anything to do with the operation of working the electric telegraph, and not a very pleasing household word I must say. I am sure most practical telegraphers would rather never hear of earth currents again. Still we have got earth currents; let us make the best of them. They are always with us; let us see whether we cannot make something out of them since they have given us so much trouble.

Now, if we could have simultaneous observations of the underground currents, of the three magnetic elements, and of the aurora, we should have a mass of evidence from which, I believe, without fail, we ought to be able to conclude an answer more or less definite to the question I have put. Are we to look in the regions external to our atmosphere for the cause of the underground currents, or are we to look under the earth for some unknown cause affecting terrestrial magnetism, and giving rise to an induction of those currents? The direction of the effects, if we can only observe those directions, will help us most materially to judge as to what answer should be given. It is my desire to make a suggestion which may reach members of this society, and associates in distant parts of the world. I make it not merely to occupy a little time in an inaugural address, but with the most earnest desire and expectation that something may be done in the direction of my suggestion. I do not venture to say that something may come from my suggestion because, perhaps, without any suggestion from me, the acute and intelligent operators whom our great submarine telegraph companies have spread far and wide over the earth are fully alive to the importance of such observations as I am now speaking of. I would just briefly say that the kind of observation which would be of value for the scientific problem is—to observe the indication of an electrometer at each end of a telegraph line at any time, whether during a magnetic storm or not, and at any time of the night or day. If the line be worked with a condenser at each end, this observation can be made without in the slightest degree influencing, and therefore without in the slightest degree disturbing, the practical work throughout the line. Put on an electrometer in direct connection with the line, connect the outside of the electrometer with a proper earth connection, and it may be observed quite irrespectively of the signalling; when the signalling is done, as it very frequently is at submarine lines, with a condenser at each end. The scientific observation will be disturbed undoubtedly, and considerably disturbed, by the sending of messages; but the disturbance is only transient, and in the very pause at the end of a word there will be a sufficiently near approach to steadiness in the potential at the end of the wire connected with the electrometer to allow a careful observer to estimate with practical accuracy the indication that he would have were there no working of the line going on at the time. A magnetic storm of considerable intensity does not stop the working, does indeed scarcely interfere with the working, of a submarine line in many instances when a condenser is used at each end. Thus, observations, even when the line is working, may be made during magnetic storms, and again, during hours when the line is not working if there are any, and even the very busiest lines have occasional hours of rest. Perhaps, then, however, the operators have no time or zeal left, or, rather, I am sure they have always zeal, but I am not sure that there is always time left, and it may be impossible for them to bear the strain longer than their office hours require them. But when there is an operator, or a superintendent, or mechanic, or an extra operator who may have a little time on his hands, then, I say, any single observation or any series of observations that he can make on the electric potentials at one end of an insulated line will give valuable results.

When arrangements can be made for simultaneous observations of the potentials by an electrometer at the two ends of the line, the results will be still more valuable. And, lastly, I may just say that when an electrometer is not available, a galvanometer of very large resistance may be employed. This will not in the slightest degree interfere with the practical working any more than would an electrometer, nor will it be more difficult to get results of the scientific observations not overpoweringly disturbed by the practical working if a galvanometer is used than when an electrometer is available. The more resistance that can be put in between the cable and the earth in circuit with a galvanometer the better, and the sensibility of the galvanometer will still be found perhaps more than necessary. Then, instead of reducing it by a shunt, let steel magnets giving a more powerful direction to the needle be applied for adjusting it. The resistance in circuit with the galvanometer between cable-end and earth ought to be at least twenty times the cable's copper-resistance to make the galvanometer observations

as valuable as those to be had by electrometer.

I should speak also of the subject of atmospheric
electricity. The electric telegraph brings this
phenomenon into connection with terrestrial magnetism, with earth currents, and, through them,
with aurora borealis, in a manner for which observations made before the time of the electric
telegraph, or without the aid of the electric telegraph, had not given us any data whatever.
Scientific observations on terrestrial magnetism,
and on the aurora, and on atmospheric electricity,
had shown a connection between the aurora and
terrestrial magnetism in the shape of the disturbances that I have alluded to at a time of magnetic
storm ; but no connection between magnetic
storms and atmospheric electricity, thunderstorms,
or generally the state of the weather—what is
commonly called meteorology—has yet been discovered. There is just one common link connecting these phenomena and those exhibited in the
electric telegraph. A telegraphic line—an air line
more particularly, but a submarine line also—

shows us unusually great disturbances not only when there are aurora and variations of terrestrial magnetism, but when the atmospheric electricity is in a disturbed state. That it should be so electricians here present will readily understand. They will understand when they consider the change of electrification of the earth's surface which a lightning discharge necessarily produces.

I fear I might occupy too much of your time, or else I would just like to say a word or two upon atmospheric electricity, and to call your attention to the quantitative relations which questions in connection with this subject bear to those of ordinary earth currents and the phenomena of terrestrial magnetism. In fair weather the surface of the earth is always, in these countries at all events, found negatively electrified. Now the limitation to these countries that I have made suggests a point for the practical telegraphists all over the world. Let us know whether it is only in England, France, and Italy that in fine weather the earth's surface is negatively electrified. The only case of exception on record to this statement is Professor Piazzi Smyth's observations on the Peak of Teneriffe. There, during several months of perfectly fair weather, the surface of the mountain was, if the electric test applied was correct, positively electrified ; but Professor Piazzi Smyth has, I believe, pointed out that the observations must not be relied upon. The instrument, as he himself found, was not satisfactory. The science of observing the atmospheric electricity was then so much in its infancy that, though he went prepared with the best instrument, and the only existing rules for using it, there was a fatal doubt as to whether the electricity was positive or negative after all. But the fact that there has been such a doubt is important.

Now I suppose there will be a telegraph to Teneriffe before long, and then I hope and trust some of the operators will find time to climb the Peak. I am sure that, even without an electric object, they will go up the Peak. Now they must go up the Peak with an electrometer in fine weather, and ascertain whether the surface of the earth is there positively or negatively electrified.

If they find that on one fine day it is negatively electrified, the result will be valuable to science ; and if on several days it is found to be all day and all night negatively electrified, then there will be a very great accession to our knowledge regarding atmospheric electricity.

When I say the surface of the earth is negatively electrified, I make a statement which I believe was due originally to Peltier. The more common form of statement is that the air is positively electrified, but this form of statement is apt to be delusive. More than that, it is most delusive in many published treatises, both in books and encyclopædias, upon the subject. I have in my mind one encyclopædia in which, in the article " Air, electricity of," it is said that the electricity of the air is positive, and increases in rising from the ground. In the same encyclopædia, in the article " Electricity, atmospheric," it is stated that the surface of the earth is negatively electrified, and that the air in contact with the earth, and for some height above the earth, is, in general, negatively electrified. I do not say

too much, then, when I say that the statement
that the air is positively electrified has been at all
events a subject for ambiguous and contradictory
propositions ; in fact, what we know by direct
observation is, that the surface of the earth is
negatively electrified, and positive electrification
of the air is merely inferential. Suppose for a
moment that there were no electricity whatever
in the air—that the air was absolutely devoid of
all electric manifestation, and that a charge of
electricity were given to the whole earth. For
this no great amount would be necessary. Such
amounts as we deal with in our great submarine
cables would, if given to the earth as a whole,
produce a very considerable electrification of its
whole surface. You all know the comparison
between the electro-static capacity of one of the
Atlantic cables, with the water round its gutta-
percha for outer coating—and the earth with air and
infinite space for its outer coating.[1] Well now, if
all space were non-conducting—and experiments

[1] The earth's radius is about 630 million centimetres, and its
electrostatic capacity is therefore 630 microfarads or about that of
1,600 miles of cable.

on vacuum tubes seem rather to support the possibility of that being the correct view—if all space were non-conducting, our atmosphere being a non-conductor, and the rarer and rarer air above us being a non-conductor, and the so-called vacuous space, or the interplanetary space beyond that (which we cannot admit to be really vacuous), being a non-conductor also, then a charge could be given to the earth as a whole, if there were the other body to come and go away again, just as a charge could be given to a pith-ball electrified in the air of this room. Then, I say, all the phenomena brought to light by atmospheric electrometers, which we observe on a fine day, would be observed just as they are. The ordinary observation of atmospheric electricity would give just the result that we obtain from it. The result that we obtain every day of fair weather in ordinary observations on atmospheric electricity is precisely the same as if the earth were electrified negatively and the air had no electricity in it whatever.

I have asserted strongly that the lower regions of the air are negatively electrified. On what foundation is this assertion made? Simply by observation. It is a matter of fact; it is not a matter of speculation. I find that the air which is drawn into a room from the outside on a fine day is negatively electrified. I believe the same phenomena will be observed in this city as in the old buildings of the University of Glasgow, in the middle of a very densely-peopled and smoky part of Glasgow; and therefore I doubt not that when air is drawn into this room from the outside, and a water-dropping collector is placed in the centre of the room, or a few feet above the floor, and put in connection with a sufficiently delicate electrometer, it will indicate negative electrification. Take an electric machine; place a spirit-lamp on its prime conductor; turn the machine for a time: take an umbrella, and agitate the air with it till the whole is well mixed up; and keep turning the machine, with the spirit-lamp burning on its prime conductor. Then apply your electric test, and you will find the air positively electrified. Again—Let two rooms, with a door and passage between them, be used for the experiment. First shut the door and open the window in your observing room. Then, whatever electric operations you may have been performing, after a short time you find indications of negative electrification of the air. During all that time, let us suppose that an electric machine has been turned in the neighbouring room, and a spirit-lamp burning on its prime conductor. Keep turning the electric machine in the neighbouring room, with the spirit-lamp as before. Make no other difference but this—shut the window and open the door. I am supposing that there is a fire in your experimenting room. When the window was open and the door closed, the fire drew its air from the window, and you got the air direct from without. Now shut the window and open the door into the next room, and gradually the electric manifestation changes. And here somebody may suggest that it is changed because of the opening of the door, and the inductive effect from the passage. But I anticipate that criticism by saying that my observation has told me that the change takes place gradually. For a time after the door is opened and the window closed, the electrification of the air in your experimental room continues negative, but it gradually becomes zero, and a little later becomes positive. It remains positive as long as you keep turning the electric machine in the other room and the door is open. If you stop turning the electric machine, then, after a considerable time, the manifestation changes once more to negative; or if you shut the door and open the window the manifestation changes more rapidly to negative. It is, then, proved beyond all doubt that the electricity which comes in at the windows of an ordinary room in town is ordinarily negative in fair weather. It is not always negative, however. I have found it positive on some days. In broken weather, rainy weather, and so on, it is sometimes positive and sometimes negative.

Now, hitherto there is no proof of positive electricity in the air at all in fine weather; but we have grounds for inferring that probably there is positive electricity in the upper regions of the air. To answer that question the direct manner is to go up in a balloon, but that takes us beyond telegraphic regions, and therefore I must stop But I do say that superintendents and telegraphic operators in various stations might sometimes make observations ; and I do hope that the companies will so arrange their work, and provide such means for their spending their spare time that each telegraph station may be a sub-section of the Society of Telegraph Engineers, and may be able to have meetings, and make experiments, and put their forces together to endeavour to arrive at the truth. If telegraph operators would repeat such experiments in various parts of the world, they would give us most valuable information. And we may hope that, besides definite information regarding atmospheric electricity, in which we are at present so very deficient, we shall also get towards that great mystery of nature— the explanation of terrestrial magnetism and its associated phenomena,—the grand circular variation of magnetism, the magnetic storms and the aurora borealis.

And now, gentlemen, I must apologise to you for having trespassed so long upon your time. I have introduced a subject which, perhaps, more properly ought to have been brought forward as a communication at one of the ordinary meetings. I may just say before sitting down, that I look forward with great hopefulness to the future of the Society of Telegraph Engineers. I look upon it as a Society for establishing harmony between theory and practice in electrical engineering—in electrical science generally. Of course, branches of engineering not purely electric are included, but the special subject of this Society is now, and I think must always be, electricity. Electric science hopes much from the observations of telegraphists, and particularly with the great means of observing that they have at their disposal. Science, I hope, will continue to confer benefits on the practical operator. Let our aim be to secure by organized co-operation that the best that science can do shall be done for the practical operator, and that the work and observations of practical operators shall be brought together through the channels of the various sub-sections, into one grand stream which this Society will be the means of utilising.

REVIEW OF EVIDENCE REGARDING THE PHYSICAL CONDITION OF THE EARTH.

[Being extract from Address to the Mathematical and Physical Section of the British Association, Glasgow, September 7th, 1876.]

THE evidence of a high internal temperature is too well known to need any quotation of particulars at present. Suffice it to say that below the uppermost ten metres stratum of rock or soil sensibly affected by diurnal and annual variations of temperature there is generally found a gradual increase of temperature downwards, approximating roughly in ordinary localities to an average rate of 1 centigrade per thirty metres of descent, but much greater in the neighbourhood of active volcanoes and certain other special localities, of comparatively small area, where hot springs and perhaps also sulphurous vapours prove an intimate relationship to volcanic quality. It is worthy of remark in passing that, so far as we know at present, there are no localities of exceptionally *small* rate of augmentation of underground temperature, and none where temperature diminishes at any time through any considerable depth downwards below the stratum sensibly influenced by summer heat and winter cold. Any considerable area of the earth of, say, not less than a kilometre in any horizontal diameter, which for several thousand years had been covered by snow or ice, and from which the ice had melted away and left an average surface temperature of 13° cent., would, during 900 years, show a decreasing temperature for some depth down from the surface ; and 3600 years after the clearing away of the ice would still show residual effect of the ancient cold, in a half rate of augmentation of temperature downwards in the upper strata, gradually increasing to the whole normal rate, which would be sensibly reached at a depth of 600 metres.

By a simple effort of geological calculus it has been estimated that 1° per 30 metres gives 1000° per 30,000 metres, and 3,333 per 100 kilometres. This arithmetical result is irrefragable ; but what of the physical conclusion drawn from it with marvellous frequency and pertinacity, that at depths of from 30 to 100 kilometres the temperatures are so high as to melt all substances composing the earth's upper crust? It has been remarked, indeed, that if observation showed any diminution or augmentation of the rate of increase of underground temperature in great depths, it would not be right to reckon on the uniform rate of 1° per 30 metres or thereabouts down to 30 or 60 or 100 kilometres. "But observation has shown nothing of the kind ; and therefore surely it is most consonant with inductive philosophy to admit no great deviation in any part of the earth's solid crust from the rate of increase proved by observation as far as the greatest depths to which we have reached!" Now I have to remark upon this argument that the greatest depth to which we have reached

in observations of underground temperature is
scarcely one kilometre ; and that if a ten per
cent. diminution of the rate of augmentation of
underground temperature downwards were found
at a depth of one kilometre, this would demonstrate [1] that within the last 100,000 years the
upper surface of the earth must have been at a
higher temperature than that now found at the
depth of one kilometre. Such a result is no
doubt to be found by observation in places
which have been overflown by lava in the memory
of man or a few thousand years further back ;
but if, without going deeper than a kilometre, a
ten per cent. diminution of the rate of increase
of temperature downwards were found for the
whole earth, it would limit the whole of geological
history to within 100,000 years, or, at all events,
would interpose an absolute barrier against the
continuous descent of life on the earth from earlier
periods than 100,000 years ago.   Therefore, al-

[1] For proof of this and following statements regarding underground heat, I refer to "Secular Cooling of the Earth," *Transactions of the Royal Society of Edinburgh*, 1862 ; and Thomson and
Tait's *Natural Philosophy*, Appendix D.

though search in particular localities for a diminution of the rate of augmentation of underground
temperature in depths of less than a kilometre
may be of intense interest, as helping us to fix
the dates of extinct volcanic actions which have
taken place within 100,000 years or so, we know
enough from thoroughly sure geological evidence
not to expect to find it, except in particular
localities, and to feel quite sure that we shall not
find it under any considerable portion of the
earth's surface. If we admit as possible any such
discontinuity within 900,000 years, we might be
prepared to find a sensible diminution of the rate
at three kilometres depth; but not at anything
less than 30 kilometres if geologists validly claim
as much as 90,000,000 of years for the length of
the time with which their science is concerned.
Now this implies a temperature of 1,000° cent. at
the depth of 30 kilometres, allows something less
than 2,000° for the temperature at 60 kilometres,
and does not require much more than 4,000° cent.
at any depth however great, but does require at
the great depths a temperature of, at all events,

not less than about 4,000° cent.   It would not
take much "hurrying up" of the actions with
which they are concerned to satisfy geologists
with the more moderate estimate of 50,000,000 of
years.   This would imply at least about 3,000°
cent for the limiting temperature at great depths.
If the actual substance of the earth, whatever it
may be, rocky or metallic, at depths of from 60
to 100 kilometres, under the pressure actually
there experienced by it, can be solid at temperatures of from 3,000° to 4,000°, then we may hold
the former estimate (90,000,000) to be as probable
as the latter (50,000,000), so far as evidence from
underground temperature can guide us.   If 4,000°
would melt the earth's substance at a depth of
100 kilometres, we must reject the former estimate though we might still admit the latter ; if
3,000° would melt the substance at a depth of
60 kilometres, we should be compelled to conclude that 50,000,000 of years is an over-estimate.
Whatever may be its age, we may be quite sure
the earth is solid in its interior ; not, I admit,
throughout its whole volume, for there certainly

are spaces in volcanic regions occupied by liquid lava ; but whatever portion of the whole mass is liquid, whether the waters of the ocean, or melted matter in the interior, these portions are small in comparison with the whole ; and we must utterly reject any geological hypothesis which, whether for explaining underground heat or ancient up-heavals and subsidences of the solid crust, or earthquakes, or existing volcanoes, assumes the solid earth to be a shell of 30, or 100, or 500, or 1,000 kilometres thickness, resting on an interior liquid mass.

If the inner boundary of the imagined rigid shell of the earth were rigorously spherical, the interior liquid could experience no precessional or nutational influence from the pressure on its bounding surface, and therefore if homogeneous could have no precession or nutation at all, or if heterogeneous only as much precession and nuta-tion as would be produced by attraction from without in virtue of non-sphericity of its surfaces of equal density, and therefore the shell would have enormously more rapid precession and nutation than it actually has—forty times as much, for instance, if the thickness of the shell is 60 kilometres. A very slight deviation of the inner surface of the shell from perfect sphericity would suffice, in virtue of the quasi-rigidity due to vortex motion, to hold back the shell from taking sensibly more precession than it would give to the liquid, and to cause the liquid (homogeneous or heterogeneous) and the shell to have sensibly the same precessional motion as if the whole constituted one rigid body. But it is only because of the very long period (26,000 years) of precession, in comparison with the period of rotation (one day), that a very slight deviation from sphericity would suffice to cause the whole to move as if it were a rigid body. A little further consideration showed me :—

(1) That an ellipticity of inner surface equal to $\frac{1}{26000 \times 865}$ would be too small, but that an ellipticity of one or two hundred times this amount would not be too small to compel approximate equality of precession throughout liquid and shell.

(2) That with an ellipticity of interior surface equal to $\frac{1}{300}$, if the precessional motion were 26,000 times as great as it is, the motion of the liquid would be very different from that of a rigid mass rigidly connected with the shell.

(3) That with the actual forces and the supposed interior ellipticity of $\frac{1}{300}$, the lunar nineteen-yearly nutation might be affected to about five per cent. of its amount by interior liquidity.

(4) Lastly, that the lunar semiannual nutation must be largely, and the lunar fortnightly nutation enormously affected by interior liquidity.

But although so much could be foreseen readily enough, I found it impossible to discover without thorough mathematical investigation what might be the characters and amounts of the deviations from a rigid body's motion which the several cases of precession and nutation contemplated would present. The investigation, limited to the case of a homogeneous liquid inclosed in an ellipsoidal shell, has brought out results which I confess have greatly surprised me. When the interior ellipticity of the shell is just too small, or the

periodic speed of the disturbance just too great
to allow the motion of the whole to be sensibly
that of a rigid body, the deviation first sensible.
renders the precessional or nutational motion of
the shell smaller than if the whole were rigid,
instead of greater, as I expected. The amount
of this difference bears the same proportion to
the actual precession or nutation as the fraction
measuring the periodic speed of the disturbance
(in terms of the period of rotation as unity) bears
to the fraction measuring the interior ellipticity
of the shell ; and it is remarkable that this result
is independent of the thickness of the shell, assumed, however, to be small in proportion to the
earth's radius.   Thus in the case of precession
the effect of interior liquidity would be to diminish the precessional motion, that is to say the
periodic speed of the precession, in the proportion stated ; in other words, it would add to the
precessional period a number of days equal to
the number whose reciprocal measures the ellipticity.   Thus, in the actual case of the earth, if we
still take $\frac{1}{300}$ as the ellipticity of the inner bound-

ary of the supposed rigid shell, the effect would be to augment by 300 days the precessional period of 2,600 years, or to diminish by about $\frac{1}{60}''$ the annual precession of about $51''$, an effect which I need not say would be wholly insensible. But on the lunar nutation of $18\cdot6$ years period, the effect of interior liquidity would be quite sensible: $18\cdot6$ years being twenty-three times 300 days, the effect would be to diminish the axes of the ellipse which the earth's pole describes in this period each by $\frac{1}{23}$ of its own amount. The semiaxes of this ellipse, calculated on the theory of perfect rigidity from the very accurately known amount of precession, and the fairly accurate knowledge which we have of the ratio of the lunar to the solar part of the precessional motion, are $9''\cdot22$ and $6''\cdot86$, with an uncertainty not amounting to one-half per cent. on account of want of perfect accuracy in the latter part of data. If the true values were less each by $\frac{1}{23}$ of its own amount, the discrepance might have escaped detection, or might *not* have escaped detection; but certainly it could be found if looked for. So far nothing can

be considered as absolutely proved with reference
to the interior solidity of the earth from precession and nutation ; but now think of the solar
semiannual and the lunar fortnightly nutations
The period of each of these is less than 300 days.
Now the hydrodynamical theory shows that, irrespectively of the thickness of the shell, the nutation of the crust would be zero if the period of
the nutational disturbance were 300 times the
period of rotation (the ellipticity being $\frac{1}{300}$) ; if
the nutational period were anything between this
and a certain smaller critical value depending
on the thickness of the crust, the nutation would
be negative ; if the period were equal to this
second critical value, the nutation would be infinite ; and if the period were still less, the nutation would be again positive.   Further, the 183
days period of the solar nutation falls so little
short of the critical 300 days that the amount
of the nutation is not sensibly influenced by the
thickness of the crust, and is negative and equal
in absolute value to $\frac{81}{59}$ (being the reciprocal of
$\frac{300}{183} - 1$) times what the amount would be were

the earth solid throughout. Now this amount, as calculated in the *Nautical Almanac,* makes o″·55 and o″·51 the semiaxes of the ellipse traced by the earth's axis round its mean position ; and if the true nutation placed the earth's axis on the opposite side of an ellipse having o″·86 and o″·81 for its semiaxes, the discrepance could not possibly have escaped detection. But, lastly, think of the lunar fortnightly nutation. Its period is $\frac{1}{20}$ of 300 days, and its amount, calculated in the *Nautical Almanac* on the theory of complete solidity, is such that the greater semiaxis of the approximately circular ellipse described by the pole is o″·0325. Were the crust infinitely thin this nutation would be negative, but its amount nineteen times that corresponding to solidity. This would make the greater semiaxis of the approximately circular ellipse described by the pole amount to 19 × o″·0885, which is 1″·7. It would be negative and of some amount between 1″·7 and infinity, if the thickness of the crust were anything from zero to 120 kilometres. This conclusion is absolutely decisive against the geological hypothesis of a thin rigid shell full of liquid.

But interesting in a dynamical point of view as Hopkins's problem is, it cannot afford a decisive argument against the earth's interior liquidity. It assumes the crust to be perfectly stiff and unyielding in its figure. This, of course, it cannot be because no material is infinitely rigid; but, composed of rock and possibly of continuous metal in the great depths, may the crust not, as a whole, be stiff enough to practically fulfil the condition of unyieldingness? No, decidedly it could not; on the contrary, were it of continuous steel and 500 kilometres thick, it would yield very nearly as much as if it were india-rubber to the deforming influences of centrifugal force and of the sun's and moon's attractions. Now although the full problem of precession and nutation, and, what is now necessarily included in it, tides, in a continuous revolving liquid spheroid, whether homogeneous or heterogeneous, has not yet been completely worked out, I think I see far enough towards a complete solution to say that precession and nutations will be practically the same in it as in a solid globe, and that the tides will be practically the same as those of the equilibrium theory. From this it follows that precession and nutations of the solid crust, with the practically perfect flexibility which it would have even though it were 100 kilometres thick and as stiff as steel, would be sensibly the same as if the whole earth from surface to centre were solid and perfectly stiff. Hence precession and nutations yield nothing to be said against such hypotheses as that of Darwin,[1] that the earth as a whole takes approximately the figure due to gravity and centrifugal force, because of the fluidity of the interior and the flexibility of the crust. But, alas for this "attractive sensational idea that a molten interior to the globe underlies a superficial crust, its surface agitated by tidal waves, and flowing freely towards any issue that may here and there be opened for its outward escape" (as Poulett Scrope called it)! the solid crust would yield so freely to the deform-

[1] "Observations on the Parallel Roads of Glen Roy and other parts of Lochaber in Scotland, with an attempt to prove that they are of marine origin." *Transactions of the Royal Society* for February 1839, p. 81.

ing influence of sun and moon that it would simply
carry the waters of the ocean up and down with it,
and there would be no sensible tidal rise and fall of
water relatively to land.

The state of the case is shortly this :—The
hypothesis of a perfectly rigid crust containing
liquid violates physics by assuming preternaturally
rigid matter and violates dynamical astronomy in
the solar semiannual and lunar fortnightly nutations ; but tidal theory has nothing to say against
it.   On the other hand, the tides decide against
any crust flexible enough to perform the nutations
correctly with a liquid interior, or as flexible as the
crust must be unless of preternaturally rigid
matter.

But now thrice to slay the slain : suppose the
earth this moment to be a thin crust of rock or
metal resting on liquid matter ; its equilibrium
would be unstable! And what of the upheavals
and subsidences? They would be strikingly analogous to those of a ship which has been rammed
—one portion of crust up and another down, and
then all down.   I may say, with some degree of

confidence, that whatever may be the relative densities of rock, solid and melted, at or about the temperature of liquefaction, it is, I think, probable that cold solid rock is denser than hot melted rock ; and no possible degree of rigidity in the crust could prevent it from breaking in pieces and sinking wholly below the liquid lava. Something like this may have gone on, and probably did go on, for thousands of years after solidification commenced —surface-portions of the melted material losing heat, freezing, sinking immediately, or growing to thicknesses of a few metres, when the surface would be cool and the whole solid dense enough to sink. " This process must go on until the sunk portions of crust build up from the bottom a sufficiently close-ribbed skeleton or frame to allow fresh incrustations to remain, bridging across the now small areas of lava pools or lakes.

" In the honeycombed solid and liquid mass thus formed there must be a continual tendency for the liquid, in consequence of its less specific gravity, to work its way up ; whether by masses of solid falling from the roofs of vesicles or tunnels and causing earthquake-shocks, or by the roof breaking quite through when very thin, so as to cause two such hollows to unite, or the liquid of any of them to flow out freely over the outer surface of the earth, or by gradual subsidence of the solid owing to the thermodynamic melting which portions of it under intense stress must experience, according to my brother's theory. The results which must follow from this tendency seem sufficiently great and various to account for all that we learn from geological evidence of earthquakes, of upheavals and subsidences of solid, and of eruptions of melted rock." [1]

Leaving altogether now the hypothesis of a hollow shell filled with liquid, we must still face the question, How much does the earth, solid through-out, except small cavities or vesicles filled with liquid, yield to the deforming (or tide-generating) influences of sun and moon? This question can only be answered by observation. A single

[1] "Secular Cooling of the Earth," *Transactions of the Royal Society of Edinburgh*, 1862 (W. Thomson), and Thomson and Tait's *Natural Philosophy*, §§ (*ee*), (*ff*).

infinitely accurate spirit-level or plummet far enough away from the sea to be not sensibly affected by the attraction of the rising and falling water would enable us to find the answer. Observe by level or plummet the changes of direction of apparent gravity relatively to an object rigidly connected with the earth, and compare these changes with what they would be were the earth perfectly rigid, according to the known masses and distances of sun and moon. The discrepance, if any is found, would show distortion of the earth and would afford data for determining the dimensions of the elliptic spheroid into which a non-rotating globular mass of the same dimensions and elasticity as the earth would be distorted by centrifugal force if set in rotation, or by tide-generating influences of sun or moon. The effect on the plumb-line of the lunar tide-generating influence is to deflect it towards or from the point of the horizon nearest to the moon, according as the moon is above or below the horizon. The effect is zero when the moon is on the horizon or overhead, and is greatest in either direction when the moon is

45° above or below the horizon. When this greatest value is reached, the plummet is drawn from its mean position through a space equal to $\frac{1}{13000000}$ of the length of the thread. No ordinary plummet or spirit-level could give any perceptible indication whatever of this effect ; and to measure its amount it would be necessary to be able to observe angles as small as $\frac{1}{13000000000}$ of the radian, or about $\frac{1}{800}''$. At present no apparatus exists within small compass by which it could be done. A submerged water-pipe of considerable length, say 12 kilometres, with its two ends turned up and open, might answer. Suppose, for example, the tube to lie north and south, and its two ends to open into two small cisterns, one of them, the southern for example, of half a decimetre diameter (to escape disturbance from capillary attraction), and the other of two or three decimetres diameter (so as to throw nearly the whole rise and fall into the smaller cistern) For simplicity, suppose the time of observation to be when the moon's declination is zero. The water in the smaller or southern cistern will rise from its lowest position to its

highest position while the moon is rising to maximum altitude, and fall again after the moon crosses
the meridian till she sets ; and it will rise and fall
again through the same range from moonset to
moonrise. If the earth were perfectly rigid, and if
the locality is in latitude 45°, the rise and fall
would be half a millimetre on each side of the
mean level, or a little short of half a millimetre
if the place is within 10° north or south of latitude
45°. If the air were so absolutely quiescent during
the observations as to give no varying differential
pressure on the two water-surfaces to the amount
of $\frac{1}{100}$ millimetre of water or $\frac{1}{1400}$ of mercury, the
observation would be satisfactorily practicable, as
it would not be difficult by aid of a microscope to
observe the rise and fall of the water in the smaller
cistern to $\frac{1}{100}$ of a millimetre ; but no such
quiescence of the atmosphere could be expected at
any time ; and it is probable that the variations of
the water-level due to difference of the barometric
pressure at the two ends would, in all ordinary
weather, quite overpower the small effect of the
lunar tide-generating motive. If, however, the two

cisterns, instead of being open to the atmosphere, were connected air-tightly by a return-pipe with no water in it, it is probable that the observation might be successfully made : but Siemens's level or some other apparatus on a similarly small scale would probably be preferable to any elaborate method of obtaining the result by aid of very long pipes laid in the ground ; and I have only called your attention to such an ideal method as leading up to the natural phenomenon of tides.

Tides in an open canal or lake of 12 kilometres length would be of just the amount which we have estimated for the cisterns connected by submerged pipe ; but would be enormously more disturbed by wind and variations of atmospheric pressure. A canal or lake of 240 kilometres length in a proper direction and in a suitable locality would give but 10 millimetres rise and fall at each end, an effect which might probably be analysed out of the much greater disturbance produced by wind and differences of barometric pressure ; but no open liquid level short of the *ingens æquor*, the ocean, will probably be found so well adapted as it for meas-

S 2

uring the absolute value of the disturbance produced on terrestrial gravity by the lunar and solar tide-generating motive. But observations of the diurnal and semi-diurnal tides in the ocean do not (as they would on smaller and quicker levels) suffice for this purpose, because their amounts differ enormously from the equilibrium-values on account of the smallness of their periods in comparison with the periods of any of the grave enough modes of free vibration of the ocean as a whole. On the other hand, the lunar fortnightly declinational and the lunar monthly elliptic and the solar semiannual and annual elliptic tides have their periods so long that their amounts must certainly be very approximately equal to the equilibrium values. But there are large annual and semi-annual changes of sea-level, probably both differential, on account of wind and differences of barometric pressure and differences of temperature of the water, and absolutely depending on rainfall and the melting away of snow and return evaporation, which altogether swamp the small semiannual and annual tides due to the sun's attraction. Happily, however, for our object, there is no meteorological or other disturbing cause which produces periodic changes of sea-level in either the fortnightly declinational or the monthly elliptic period ; and the lunar gravitational tides in these periods are therefore to be carefully investigated in order that we may obtain the answer to the interesting question, How much does the earth as an elastic spheroid yield to the tide-generating influence of sun or moon ?   Hitherto in the British Association Committee's reductions of Tidal Observations we have not succeeded in obtaining any trustworthy indications of either of these tides.   The St. George's Pier landing-stage pontoon, unhappily chosen for the Liverpool tide-gauge, cannot be trusted for such a delicate investigation: the available funds for calculation were expended before the long-period tides for Hilbre Island could be attacked, and three years of Kurrachee gave our only approach to a result. Comparisons of this with an indication of a result of calculations on West Hartlepool tides, conducted with the assistance of a grant from the

Royal Society seem to show possibly no sensible yielding, or perhaps more probably some degree of yielding, of the earth's figure. The absence from all the results of any indication of a 18·6 yearly tide (according to the same law as the other long-period tides) is not easily explained without assuming or admitting a considerable degree of yielding.

Closely connected with the question of the earth's rigidity, and of as great scientific interest and of even greater practical moment, is the question, How nearly accurate is the earth as a timekeeper? and another of, at all events, equal scientific interest, How about the permanence of the earth's axis of rotation?

Peters and Maxwell, about 35 and 25 years ago, separately raised the question, How much does the earth s axis of rotation deviate from being a principal axis of inertia? and pointed out that an answer to this question is to be obtained by looking for a variation in latitude of any or every place on the earth's surface in a period of 306 days. The model before you illustrates the

travelling round of the instantaneous axis relatively to the earth in an approximately circular
cone whose axis is the principal axis of inertia,
and relatively to space in a cone round a fixed
axis.   In the model the former of these cones,
fixed relatively to the earth, rolls internally on the
latter supposed to be fixed in space.   Peters gave
a minute investigation of observations at Pulkova
in the years 1841-42, which seem to indicate at
that time a deviation amounting to about $\frac{3}{10}''$ of
the axis of rotation from the principal axis.   Maxwell, from Greenwich observations of the years
1851-54, found seeming indications of a very slight
deviation, something less than half a second, but
differing altogether in phase from that which the
deviation indicated by Peters, if real and permanent, would have produced at Maxwell's later time.
On my begging Professor Newcomb to take up the
subject, he kindly did so at once, and undertook to
analyse a series of observations suitable for the
purpose which had been made in the United States
Naval Observatory, Washington.   A few weeks
later I received from him a letter referring me to a

paper by Dr. Nysen, of Pulkova Observatory, in which a similar negative conclusion as to constancy of magnitude or direction in the deviation sought for is arrived at from several series of the Pulkova observations between the years 1842 and 1872, and containing the following statement of his conclusions [1] :—

" The investigation of the ten-month period of latitude from the Washington prime vertical observations from 1862 to 1867 is completed, indicating a coefficient too small to be measured with certainty. The declinations with this instrument are subject to an annual period which made it necessary to discuss those of each month separately. As the series extended through a full five years, each month thus fell on five nearly equidistant points of the period. If x and y represent the coordinates of the axis of instantaneous rotation on June 30, 1864, then the observations of the separate months give the following values of x and y :—

---

[1] [For later investigations by Newcomb and others on this subject see extracts from my Presidential Addresses of 1891 and 1892 to the Royal Society.—K., March 22, 1893.]

	x.	Weight.	*v.*	Weight.
January	−0ʺ·35	10	+0ʺ·32	
February . . .	−0·03	14	+0·09	
March	+0·17	10	+0·16	
April	+0·44	5	+0·05	
May	+0·08	16	+0·02	
June	−0·01	14	−0·01	
July	−0·05	14	0·00	
August	−0·24	14	+0·29	
September . .	+0·18	14	+0·21	
October	+0·13	14	−0·01	
November . .	+0·08	17	−0·20	
December . .	−0·08	16	−0·08	
Mean	0·01 ± 0·03		+0·05 ± 0·03	

"Accepting these results as real, they would indicate a radius of rotation of the instantaneous axis amounting, at the earth's surface, to 5 feet and a longitude of the point in which this axis intersects the earth's surface near the North Pole, such that on July 11, 1864, it was 180° from Washington, or 103° east of Greenwich. The excess of the coefficient over its probable error is so slight that this result cannot be accepted as anything more than a consequence of the unavoidable errors of observation."

From the discordant character of these results

we must not, however, infer that the deviations indicated by Peters, Maxwell, and Newcomb are unreal. On the contrary, any that fall within the limits of probable error of the observations ought properly to be regarded as real. There is, in fact, a *vera causa* in the temporary changes of sea-level due to meteorological causes, chiefly winds, and to meltings of ice in the polar regions and return evaporations, which seems amply sufficient to account for irregular deviations of from $\frac{1}{2}''$ to $\frac{1}{30}''$ of the earth's instantaneous axis from the axis of maximum inertia, or, as I ought rather to say, of the axis of maximum inertia from the instantaneous axis.

As for geological upheavals and subsidences, if on a very large scale of area, they must produce, on the period and axis of the earth's rotation, effects comparable with those produced by changes of sea-level equal to them in vertical amount. For simplicity, calculating as if the earth were of equal density throughout, I find that an upheaval of all the earth's surface in north latitude and east longitude and south latitude and west longitude with equal depression in the other two quarters, amounting at greatest to 10 centimetres, and graduating regularly from the points of maximum elevation to the points of maximum depression in the middles of the four quarters, would shift the earth's axis of maximum moment of inertia through 1″ on the north side towards the meridian of 90° W. longitude, and on the south side towards the meridian of 90° E. longitude. If such a change were to take place suddenly, the earth's instantaneous axis would experience a sudden shifting of but $\frac{1}{300}$″ (which we may neglect), and then, relatively to the earth, would commence travelling, in a period of 306 days, round the fresh axis of maximum moment of inertia. The sea would be set into vibration, one ocean up and another down through a few centimetres, like water in a bath set aswing. The period of these vibrations would be from 12· to 24 hours, or at most a day or two; their subsidence would probably be so rapid that after at most a few months they would become insensible. Then a regular 306-days period tide of 11 centimetres from lowest to highest would be to be observed, with gradually diminishing amount from century to century, as through the dissipation of energy produced by this tide the instantaneous axis of the earth is gradually brought into coincidence with the fresh axis of maximum moment of inertia. If we multiply these figures by 3600, we find what would be the result of a similar sudden upheaval and subsidence of the earth to the extent of 360 metres above and below previous levels. It is not impossible that in the very early ages of geological history such an action as this, and the consequent 400-metres tide producing a succession of deluges every 306 days for many years, may have taken place ; but it seems more probable that even in the most ancient times of geological history the great world-wide changes, such as the upheavals of the continents and subsidences of the ocean-beds from the general level of their supposed molten origin, took place gradually through the thermodynamic melting of solids and the squeezing out of liquid lava from the interior, to which I have already referred. A slow

distortion of the earth as a whole would never produce any great angular separation between the instantaneous axis and the axis of maximum moment
of inertia for the time being. Considering, then,
the great facts of the Himalayas and Andes, and
Africa and the depths of the Atlantic, and America
and the depths of the Pacific, and Australia, and
considering further the ellipticity of the equatorial
section of the sea-level estimated by Capt. Clarke
at about $\frac{1}{10}$ of the mean ellipticity of meridional
sections of the sea-level, we need no brush from
the comet's tail (a wholly chimerical cause which
can never have been put forward seriously except
in ignorance of elementary dynamical principles)
to account for a change in the earth's axis ; we
need no violent convulsion producing a sudden
distortion on a great scale, with change of the
axis of maximum moment of inertia followed by
gigantic deluges ; and we may not merely admit,
but assert as highly probable, that the axis of
maximum inertia and axis of rotation, always very
near one another, may have been in ancient times
very far from their present geographical position

and may have gradually shifted through 10, 20, 30, 40, or more degrees without at any time any perceptible sudden disturbance of either land or water.

Lastly as to variations in the earth's rotational period. You all no doubt know how, in 1853, Adams discovered a correction to be needed in the theoretical calculation with which Laplace followed up his brilliant discovery of the dynamical explanation of an apparent acceleration of the moon's mean motion shown by records of ancient eclipses, and how he found that when his correction was applied the dynamical theory of the moon's motion accounted for only about half of the observed apparent acceleration, and how Delaunay in 1866 verified Adams's result and suggested that the explanation may be a retardation of the earth's rotation by tidal friction. The conclusion is that, since the 19th of March, 721 B.C., a day on which an eclipse of the moon was seen in Babylon, commencing "when one hour after her rising was fully passed," the earth has lost rather more than $\frac{1}{3,000,000}$ of her rotational velocity, or, as a timekeeper, is going slower by $11\frac{1}{2}$ seconds per annum now than then.

According to this rate of retardation, if uniform, the earth at the end of a century would, as a timekeeper, be found 22 seconds behind a perfect clock, rated and set to agree with her at the beginning of the century. Newcomb's subsequent investigations in the lunar theory have on the whole tended to confirm this result ; but they have also brought to light some remarkable apparent irregularities in the moon s motion, which, if real, refuse to be accounted for by the gravitational theory without the influence of some unseen body or bodies passing near enough to the moon to influence her mean motion. This hypothesis Newcomb considers not so probable as that the apparent irregularities of the moon are not real, and are to be accounted for by irregularities in the earth's rotational velocity. If this is the true explanation, it seems that the earth was going slow from 1850 to 1862, so much as to have got behind by 7 seconds in these 12 years, and then to have begun going faster again so as to gain 8 seconds from 1862 to 1872. So great an irregularity as this would require somewhat greater changes of sea-level, but not many times greater than the British Association Committee's reductions of tidal observations for several places in different parts of the world allow us to admit to have possibly taken place. The assumption of a fluid interior, which Newcomb suggests, and the flow of a large mass of the fluid " from equatorial regions to a position nearer the axis," is not, from what I have said to you, admissible as a probable explanation of the remarkable acceleration of rotational velocity which seems to have taken place about 1862 ; but happily it is not necessary. A settlement of 14 centimetres in the equatorial regions, with corresponding rise of 28 centimetres at the poles (which is so slight as to be absolutely undiscoverable in astronomical observatories, and which would involve no change of sea-level absolutely disproved by reductions of tidal observations hitherto made), would suffice. Such settlements must occur from time to time ; and a settlement of the amount suggested might result from the diminution of centrifugal force due to 150 or 200 centuries tidal retardation of the earth's rotational speed.

GEOLOGICAL CLIMATE.

[*Being a Paper read before the Geological Society of Glasgow,*
February 22, 1877.]

THE subject on which I am to speak to you this evening, is one which has interested and exercised geologists from the very beginning of their science. We find in geological strata many evidences of differences of climate through different periods of past time ; and strenuous endeavours have been made to account for those differences. The subject has acquired a special interest within the last few months, through the return of the Arctic expedition bringing evidences of a very warm climate, probably as warm as we have it now in the tropics, within nine degrees of the North Pole. It had been well known that places far north as for instance Melville Island and Prince Patrick's Island, within fifteen degrees of the pole, had at some time, not very remote, enjoyed a climate comparable with our own or at all events comparable with the climates of Norway, Sweden, and the north of Scotland. On this point I find a most interesting statement in Croll's book, *Climate and Time.* Although it is no doubt well known to many present I am sure all will hear it with interest if I read it to you. It is entitled "Evidence of Warm Periods in Arctic Regions."—"The fact that stumps, etc., of full-grown trees have been found in places where at present nothing is to be met with but fields of snow and ice, and where the mean annual temperature scarcely rises above the zero of the Fahrenheit Thermometer, is good evidence to show that the climate of the arctic regions was once much warmer than now. The remains of an ancient forest were discovered by Captain M'Clure, in Banks's Land, in latitude 74° 48'. He found a great accumulation of trees, from the sea level to an elevation of upwards of 300 feet. 'I entered a ravine,' says Captain M'Clure, 'some miles inland, and found the north side of it, for a depth of 40 feet from the surface, composed of one mass of wood similar to what I had before seen.'[1] In the ravine he observed a tree protruding about 8 feet, and 3 feet in circumference. And he further states that, *From the perfect state of the bark*, and the position of the trees so far from the sea, there can be but little doubt that they grew originally in the country.' A cone of one of the fir trees was brought home, and was found to belong apparently to the genus *Abies*, resembling *A. (Pinus) alba.*

" In Prince Patrick's Island, in latitude 76° 12' N., longitude 122° W., near the head of Walker Inlet, and a considerable distance in the interior in one of the ravines, a tree, protruding about 10 feet from the bank was discovered by Lieutenant Mecham. It proved to be 4 feet in circumference In its neighbourhood several others were seen, all of them similar to some he had found at Cape Manning ; each of them measured 4 feet round and 30 feet in length. The carpenter stated that the trees resembled larch. Lieutenant Mecham,

[1] *Discovery of the North- West Passage*, page 213.

from their appearance and position, concluded that they must have grown in the country.[1]

" Trees under similar conditions were also found by Lieutenant Pim on Prince Patrick's Island, and by Captain Parry on Melville Island, all considerably above the present sea-level and at a distance from the shore. On the coast of New Siberia, Lieutenant Anjou found a cliff of clay containing stems of trees still capable of being used for fuel.

" 'This remarkable phenomenon,' says Captain Osborne, 'opens a vast field for conjecture, and the imagination becomes bewildered in trying to realise that period of the world's history when the absence of ice and a milder climate allowed forest trees to grow in a region where now the ground-willow and the dwarf-birch have to struggle for existence.'

* * * "All who have seen those trees in arctic regions agree in thinking that they grew *in situ.* And Professor Haughton, in his excellent account of the arctic archipelago appended to

[1] *Voyage of the Resolute*, page 294.

M'Clintock's *Narrative of Arctic Discoveries*, after a careful examination of the entire evidence on the subject is distinctly of the same opinion ; while the recent researches of Professor Heer put it beyond doubt that the drift theory must be abandoned.

"But in reality we are not left to theorise on the subject, for we have a well authenticated case of one of those trees being got by Captain Belcher standing erect in the position in which it grew. It was found immediately to the northward of the narrow strait opening into Wellington Sound, in latitude 75° 32' N., longitude 92° W., and about a mile and a half inland. The tree was dug up out of the frozen ground, and along with it a portion of the soil which was immediately in contact with the roots. The whole was packed in canvas and brought to England. Near to the spot several knolls of peat mosses about nine inches in depth were found, containing the bones of the lemming in great numbers. The tree in question was examined by Sir William Hooker, who gave the following report concerning it, which bears out strongly the fact of its having grown *in situ.*

"'The piece of wood brought by Sir Edward Belcher from the shores of Wellington Channel belongs to a species of pine probably the *Pinus (Abies) alba*, the most northern conifer.[1] The structure of the wood of the specimen brought home differs remarkably in its anatomical character from that of any other conifer with which I am acquainted. Each concentric ring (or annual growth) consists of two zones of tissue ; one, the outer, that towards the circumference, is broader, of a pale colour and consists of ordinary tubes of fibres of wood marked with discs common to all coniferæ These discs are usually opposite one another when more than one row of them occur in the direction of the length of the fibre ; and, what is very unusual, present radiating lines from the central depression to the circumference. Secondly, the inner zone of each annual ring of wood is narrower, of a dark colour, and formed of more slender woody fibres, with thicker walls in proportion to

[1] *Quarterly Journal Geological Society,* Vol. XI., page 540.

their diameter. These tubes have few or no discs upon them but are covered with spiral striæ, giving the appearance of each tube being formed of a twisted band. The above characters prevail in all parts of the wood, but are slightly modified in different rings. Thus the outer zone is broader in some than in others, the disc bearing fibres of the outer zone are sometimes faintly marked with spiral striæ ; and the spirally marked fibres of the inner zone sometimes bear discs. These appearances suggest the annual recurrence of some special cause that shall thus modify the first and last formed fibres of each year's deposit, so that that first formed may differ in amount as well as in kind from that last formed ; and the peculiar conditions of an arctic climate appear to afford an adequate solution. The inner or first-formed zone, must be regarded as imperfectly developed, being deposited at a season when the functions of the plant are very intermittently exercised, and when a few short hours of sunshine are daily succeeded by many of extreme cold. As the season advances the sun's heat and light are continuous during the greater part of the twenty-four hours, and the newly formed wood fibres are hence more perfectly developed, they are much longer, present no signs of striæ, but are studded with discs of a more highly organized structure than are usual in the natural order to which this tree belongs.' [1]

" Another circumstance which shows that the tree had grown where it was found is the fact that in digging up the roots portions of the leaves were obtained. It may also be mentioned that near this place was found an old river channel cut deeply into the rock, which at some remote period, when the climate must have been less rigorous than at present, had been occupied by a river of considerable size."

The theory set forth bv Lyell in the celebrated twelfth chapter of his *Principles of Geology*, according to which variations of climate are explained by variations in the distribution of land and sea, seems amply sufficient to account for forests of such trees as now flourish in northern

[1] *British Association Report for* 1885, page 381. *The Last of the Arctic Voyages*, Vol. I., page 381

Europe having grown over those already explored arctic islands on the north side of North America where we have found their remains, and to lead us to expect that similar remains of ancient forests may be found anywhere within the arctic circle where land is reached and explored in future, even if up to the very North Pole. Just look at this map of the circumpolar regions. Look at this great north polar Mediterranean sea, everywhere about 2000 miles across. At present there is just one *wide* entrance to it from temperate waters, that "Greenland sea," a channel 720 miles broad between Norway and East Greenland opening the Polar Sea to the North Atlantic. On the west side of Greenland, between it on the east and Ellesmere Land and Grinnel Land on the west, there is another entrance—the long narrow channel of Smith's Sound, Kennedy Channel, and Robeson Channel, 600 miles long, narrowing to about 30 miles in several places, and to 20 miles in the very portal of the Polar Sea at the north end of Robeson Channel. There are besides entrances by long, intricate, narrow channels among the

islands of the North American Archipelago west
of Ellesmere Land and Grinnel Land : and just
one other entrance—that from the North Pacific,
Behring's Strait 50 miles broad. Travel now
along the boundary of the North Polar Sea east
wards ; first along the north coast of Europe and
Asia, from Norway round to Cape North in the
far east of Siberia ; thence on eastwards in west
longitude along the north coast of the extreme
eastern peninsula of Siberia ; then across Behring's
Strait along the north coast of the north American
continent to Mackenzie River and Franklin Bay :
then northwards and westwards along the eastern
and northern coasts of the outlying islands of that
great north American Archipalego so splendidly
explored by British Arctic expeditions within the
last 60 years : then from Parry Islands and
Queen's Channel and Belcher Channel strike
northwards and explore 300 miles of unknown
but probably existing east and north east coasts of
Ellesmere Land and Grinnel land till you come to
Cape Alfred Ernest and Point Alert. Travelling
thence eastwards on Lieut. Aldrich's sledge route

of 225 miles along the north coast of Grinnel
Land, you find yourself once more among familiar
names ;[1] Cape Evans, Milne Bay, North-west
Cape May, Cape Richards, Cape Stephenson, Cape
Alexandra, Ward Hunt Island, Disraeli Bay, Cape
Albert Edward, Cape Nares, Cape Aldrich, Parr
Bay Clements Markham Inlet, Feilden Peninsula,
Cape Joseph Henry, Paterson Bay, Hilgard Bay,
Cape Belknap ; till you come to the Alert's winter
quarters of 1875–6. Then strike across the mouth
of Robeson Channel 20 miles and you find 120
miles of the North coast of Greenland, well
surveyed by Lieut. Beaumont. Beyond the end
of his sledge route there are notable features,
Mount Farragut, Mount Coppinger, North-east
Cape May, Mount Hooker, Beaumont Island, with
its Mount Albert ; and at a distance of 30 miles
Cape Britannia ; all sketched and their positions
accurately fixed by "angles." Lithographed views
of them are published in the Admiralty Blue-book,

[1] These names are all to be found in the large-scale charts of the
Admiralty Blue Book containing the full report of the recent Arctic
Expedition

so that you will recognize them when you see them, as readily and surely as a stranger entering the British Channel can recognize St. Michael's Mount, or the Lizard, or Rame Head, or Portland, or Bolt Head, or the Isle of Wight. From Cape Britannia you have 500 miles of unknown but probably existing north and north-east coast line of Greenland to travel over till you find yourself at Cape Bismark : thence a few miles south and you may rest in Ardencaple Inlet having completed your 6300 miles journey round the whole coast of the Arctic sea. But that unexplored 300 miles of Ellesmere Land and Grinnel Land from Belcher Channel to Cape Alfred Ernest, and that 500 miles of North and East Greenland ! Surely Austria, Germany, America, and England will continue their great work of Arctic exploration : and may we not hope that happier times will soon come, when France and Russia will join in the amicable rivalry—not for the mere object, or perhaps not at all for the object, of reaching the north pole, but to explore the still unexplored 800 miles of coast line of the Arctic

Sea, to find how far west and north Kellet's Land extends, and to discover all that can be discovered of the Austrian Islands north east of Spitzbergen.

Now look once more at the arctic chart, and imagine the circumpolar land to be one or two thousand feet lower relatively to the sea level than at present. Such an effort of imagination does not take away your breath. You are well trained for it by the succession of elevations and subsidences which you have been called on to admit for explanation of the familiar features of our Scottish hills and glens. A thousand feet of depression would submerge the continents of Europe, Asia, and America, for thousands of miles from their present northern coast lines ; and would give instead of the present land-locked and therefore ice-bound Arctic sea, an open iceless ocean, with only a number of small steep islands to obstruct the free interchange of water between the North Pole and temperate or tropical regions. That the arctic sea would, in such circumstances, be free from ice quite up to the north pole may be I think securely inferred from what in the present condition of the globe, we know of ice-bound and open seas in the northern hemisphere and of the southern ocean abounding in icebergs but probably nowhere ice-bound up to the very coast of the circumpolar Antarctic continent, except in more or less land-locked bays. We can scarcely doubt, from what we actually see, that if that Antarctic continent were not there to receive snow (or hoar frost) to let it be kneaded into ice-sheet and glaciers, and to deliver masses of ice to be broken and washed away from its coasts, there would be neither ice nor icebergs in the southern hemisphere. Frozen seas, and icebergs in open seas, could not exist without land or very shallow water, in polar regions. Suppose sun and seasons, and climate, and distribution of sea and land to be all just as they are but with this difference that the Antarctic fixed ice be not supported by land above the sea level as it is, but be aground on solid earth everywhere 100 fathoms below the sea-level. The undermining influence of the water washing against the ice all round would certainly cause more of icebergs to break away from it than at present : its area would therefore diminish (for it does not increase at present). The smaller the area of ice round the pole, the more rapidly must its boundary be eaten into by the water, and the less abundant can be the replenishment from the interior towards compensating for this consumption. Hence the ice must all melt away in time. Suppose now the sea, unobstructed by land from either pole to temperate or tropical regions, to be iceless at any time, would it continue iceless during the whole of the sunless polar winter? *Yes* we may safely answer. Supposing the depth of the sea to be not less than 50 or 100 fathoms, and judging from what we know for certain of ocean currents, we may safely say that differences of specific gravity of the water produced by difference of temperature not reaching anywhere down to the freezing point, would cause enough of circulation of water between the polar and temperate or tropical regions to supply all the heat radiated from the water within the arctic circle during the sunless winter, if air contributed none of it. Just think of a current of three quarters of a nautical mile per

288 POPULAR LECTURES AND ADDRESSES.

hour or 70 miles per four days, flowing towards the pole across the arctic circle. The area of the arctic circle [1] is 700 square miles for each mile of its circumference. Hence 40 fathoms deep of such a current would carry in, per twenty-four hours, a little more than water enough to cover the whole area to a depth of 1 fathom: and this, if $7°·1$ Cent. above the freezing point would bring in just enough of heat to prevent freezing, if in twenty-four hours as much heat were radiated away as taken from a tenth of a fathom of ice-cold water would leave it ice at the freezing point. This is no doubt much more than the actual amount of radiation and the supposed current is probably much less than would be if the water were ice-cold at the pole and $7°$ Cent. at the Arctic circle. Hence, without any assistance from air, we find in the convection of heat by water alone a sufficiently powerful influence to prevent any freezing up in polar regions at any time of year. But the air also must carry much heat— quantities of heat considerable in comparison with

[1] Being approximately plane, and of 1400 miles radius.

those carried by ocean currents, from temperate to polar regions. Think of a south wind of 15 nautical miles per hour through a space 2300 feet above the surface. The air thus in motion will be $\frac{1}{11}$ of the whole of the atmosphere over the region, and will therefore be equal in weight to 3 fathoms of water and equal in thermal capacity to $\frac{3}{4}$ of a fathom of water. This, if at the same temperature as our previously supposed current of water of three quarters of a mile per hour will carry as much heat as 15 fathoms deep of the water current; without taking into account the latent heat of aqueous vapour condensed into liquid water in the air as it cools, which gives a large addition to the heat carrying capacity of wind in ordinary atmospheric conditions. For example air at 15° Cent. saturated with aqueous vapour must have twice as much heat taken from it as dry air at the same temperature, to cool it to 0° Cent.

According to my brother Prof. James Thomson's theory of the great primary and secondary circulations of the atmosphere, the earth's surface in the temperate and circumpolar regions must experience a prevalence of South-west winds in the Northern hemisphere, and North-west winds in the Southern hemisphere. The thermal estimates which we have just had under consideration show that these poleward winds must in all possible conditions as to distribution of land and water, have a great direct influence in moderating the cold of the northern regions. They contribute also largely towards the same result by increasing the oceanic circulation, and would do so to a still greater degree if there were less of land than there is in the circumpolar and temperate regions.

Now what must be the climate of a small island in an open iceless circumpolar sea ? Temperate and quite free from frost except in hollows, we may I think safely answer. The water and air all round it are above the freezing point, and the air is saturated or nearly saturated with aqueous vapour Wells' old theory of dew now helps us. When the sun is below the horizon the upper surface of the earth whether it be rock or stones or soil, or blades of grass or leaves of trees whatever is unscreened from above, radiates as long as its temperature is higher than the average radiational temperature of the *atmosphere* and *space* above it. This temperature will be the temperature of the under side of the cloudy stratum, and will generally be above the freezing point if the atmosphere is thickly clouded : it will be perhaps not much above the temperature of planetary space[1] when the air is at its clearest. What then is there to prevent fine leaves of trees or grass in any part of the world from destruction by cooling down to near the temperature of space, in ten minutes after the sun sets ? Answer :— *clouds*, or if not clouds, *wind;* or if neither clouds nor wind, *dew* and *ground mist;* clouds by moderating or annulling the radiation : wind by supplying heat from so considerable a mass of air that no part of it is cooled to the dew point : dew by the latent heat of the vapour from which it is formed when clouds and wind do not suffice to prevent the temperature from sinking to the dew point, and ground mist (or dew in the motionless stratum

[1] That is to say the temperature shown by a thermometer held in space on the night side of the earth a few thousand miles from it.

of air next the ground) by partially screening
the ground from radiation and itself instead
radiating heat which it draws from a thicker
stratum of still air than could contribute sensibly
by conduction to supply radiation from the
ground. If these defences suffice to protect
vegetation against destructive cold for a summer
night over a flat continent in tropical or temperate
regions it seems certain that they would suffice
to prevent even so much as hoar frost on a small
island at the very pole during its whole winter
six months night, if it were surrounded by a deep
ocean with no land to obstruct free circulation
between it and tropical seas.

Considering, then, the absolute proofs we have
in Geology of subsidences and elevations of vast
regions of the solid earth, I see no difficulty in
admitting the complete sufficiency of the theory
set forth in Lyell's twelfth chapter to account
for all that is proved of temperate climates in
polar regions, and of alternations of glacial and
mild climates in Europe, and North America,
and New Zealand. And if a few scratched rocks

prove (and there seems fair reason to believe they prove) a glacial climate to have prevailed at one time in some part or parts of India, surely the 15000 feet elevation of a district required to account for it in that way is a smaller assumption than the 35000 feet diminution of distance from the earth's centre to the surface for thousands of miles round the locality needed to bring the pole to that region.

But the astronomical cause invoked by Herschel must have had, and must now have, its effect. It does not go for nothing that we are nearer the sun by one thirtieth of the mean distance in the midwinter than in the midsummer of the northern hemisphere, and in the midsummer than in the midwinter of the southern hemisphere. Eleven thousand five hundred years ago this was reversed:[1] and if the distribution of land and water were then the same or nearly the same in both hemispheres as at present, there must have been more difference between winter and summer in the northern hemi-

Through the precession of the equinoxes and the secular variation of the position of the major axis of the earth's orbit.

sphere and less in the southern then than now. And whatever effect on the seasons or climates of the two hemispheres is produced by the change in a few thousand years from being *nearest to the sun in winter* to *being nearest to the sun in summer*, much greater effect must have been produced when through the secular variation of the eccentricity of the earth's orbit the difference of distance has been three or four times as great as at present, as it has been probably more than once in the last million years. But what the precise character of this effect may be, or whether even at its greatest it can have been distinctly sensible among the undoubtedly potent influences depending on the distribution of land and sea, or whether it can have been preponderant in giving rise to alternations of glacial and milder epochs as argued by Mr. Croll in the interesting and suggestive speculations of his *Climate and Time*, are very difficult questions which have not hitherto been validly dealt with and which deserve the most careful consideration.

As to changes of the earth's axis, I need not

repeat the statement of dynamical principles which I .gave with experimental illustrations to the Society three years ago ; but may remind you of the chief result which is that, for steady rotation, the axis round which the earth revolves must be a " principal axis of inertia," that is to say such an axis that the centrifugal forces called into play by the rotation balance one another. The vast. transpositions of matter at the earth's surface or else distortions of the whole solid mass, which must have taken place to alter the axis sufficiently to produce sensible change of the climate in any region must be considered and shown to be possible or probable before any hypothesis accounting for changes of climate by alterations of the axis can be admitted. This question has been exhaustively dealt with by Mr. George Darwin in a paper recently communicated to the Royal Society of London, and the requisitions of dynamical mathematics for an alteration of even as much as two or three degrees in the earth's axis in what may be practically called geological time shown to be on purely geological grounds exceedingly improbable. But even suppose such a change as would bring ten or twenty degrees of more indulgent sky to the American Arctic Archipelago;—it would bring Nova Zembla and Siberia by so much nearer to the pole: and it seems that there is probably as much need of accounting for a warm climate on one side of the pole as on the other. There is in fact no evidence in geological climate throughout those parts of the world which geological investigation has reached, to give any indication of the poles having been anywhere but where they are, at any period of geological time.

Pass now from the evidence of a temperate climate, allowing vegetation such as at present lives in the North of Europe, to have lived all over the Arctic islands at a not very remote geological epoch, of which we have found a perfectly satisfactory explanation in Lyell's hypothesis of the submergence of the circumpolar land barriers ; and consider next the evidence, with which geological investigation teems, of warmer climate all over the earth from equator to poles in the more ancient geological periods Underground heat, though certainly greater in the earlier geological times, cannot, as I have shown elsewhere,[1] have ever sensibly influenced the climate. Ten, twenty, thirty times the present rate of augmentation of temperature downwards could not raise the surface temperature of the earth and air in contact with it by more than a small fraction of a degree Fahrenheit. The earth might be a globe of white hot iron covered with a crust of rock 2000 feet, or there might be an ice-cold temperature everywhere within 50 feet of the surface, yet the climate could not on that account be sensibly different from what it is, or the soil be sensibly more or less genial than it is for the roots of trees or smaller plants. Yet greater underground heat is the hypothesis which has been most complacently dealt with by geologists to account for the warmer climates of ancient times.

---

[1] *Secular Cooling of the Earth,* Trans. R.S.E., 1862, and Thomson and Tait's *Natural Philosophy,* Vol. I., App. D.

The one hypothesis of all hitherto suggested that has received no favour from any professed geologist is that of a warmer sun—the one hypothesis that is rendered almost infinitely probable by independent physical evidence and mathematical calculation.

THE INTERNAL CONDITION OF THE EARTH; AS TO TEMPERATURE, FLUIDITY, AND RIGIDITY.

[Being Paper read before the Geological Society of Glasgow, February 14, 1878.]

ON previous occasions I have referred to the various arguments that have been adduced with reference to the interior condition of the earth, and to the different grounds on which it may be concluded that, instead of being a liquid mass enclosed by a mere thin shell of solid material, the earth must be, on the whole, solid.

That there is some liquid within the earth is not to be denied. Immediately before lava breaks forth in the eruption of a volcano there is liquid in the interior ; and, from volcanic eruptions, we know that there is sometimes, and that there may be always, some liquid in the earth's interior. How much liquid there may be—how much it is necessary to assume in order to account for the known phenomena of volcanic eruptions—cannot be estimated with any degree of definiteness; but I hope to be able to put before you, this evening, arguments which will convince you that we cannot admit that there is any great liquid ocean under the earth's surface, and that we are forced to look to local causes for the explanation of volcanic eruptions.

The main reason given for supposing the interior of the earth to be fluid is not, however, the existence of volcanic phenomena. From considerations regarding underground temperature, many geologists have been led to hold as a geological truth that the interior of the earth is molten throughout.

The evidences of a high internal temperature are well known. It is found by observation that passing downwards from the surface, we meet first with a thin stratum in which variations of temperature, due to the conduction inward of summer heat and winter cold, are perceptible. This stratum is on the average about 10 metres thick. Its thickness, however, is different in different localities. It depends on the thermal conductivity of the rocks composing it. At greater depths than about 10 metres the effect of the variations of external temperature with day and night and with the seasons is not sensible. As we go down lower and lower we find that the temperature of the strata increases steadily. Speaking roughly, we find that, on the average, the increase of underground temperature in ordinary localities is about 1° C. per 30 metres of descent, or about 1° F. per 50 feet.

There are localities, such as Kreuznach in Rhenish Prussia, in which the rate of increase of underground temperature, as we proceed downwards, is much greater than this. These are places, no doubt, where disturbances, in the way of great outbursts of lava, have taken place in comparatively recent times ; and I may suggest, that careful observation of the rate of increase of underground temperature, in places where the date is known of a recent disturbance, would be of great importance, and would give us the means of inferring something regarding the geological history of places where an extraordinary rate of increase is observed. It is not, however, my purpose, at present, to enter into any speculation regarding geological dates in cases such as these, but rather to put before you considerations with respect to underground temperatures in general, which in the first place seem to indicate internal fluidity, but which, when further studied, do not really lead us to any such conclusion, but rather lead us to suspend our judgment, and to look in other directions for evidences of the internal condition of the earth as to whether it is fluid or solid.

From the observed increase of temperature downwards below ground in all localities, we must conclude that the earth had, at some time more or less remote, its whole surface at some very high temperature. It may have been red-hot and solid all round ; but it is far more probable that the surface, at least, was molten all over ; and it is quite likely that the hypothesis more commonly held—namely, that it was molten throughout—is true. The most probable hypothesis that has been given for the early history of the earth is, that it was built up by the falling in of meteoric matter, and that the matter falling in with great velocity, acquired through gravitation, became heated by collision. If this hypothesis is true, the heat caused by the collisions, after the earth had grown to be anything approaching to its present size, would be far more than sufficient to melt the material falling in.

Let us now consider a red-hot liquid globe, and suppose it to cool as does lava. The result of the cooling is that as soon as any portion of the matter reaches a temperature below the melting point, it freezes and becomes solid. Instead of molten lava, it becomes solid rock. And now comes the great question—will the solid rock thus formed sink or float ? I wish this question could be answered. We have at present very little information on the subject—none of a definite kind, so far as I know, except that given us by the experiments of Bischoff.[1] Bischoff, experimenting on trachite, granite, and basalt, found the solids in these three cases about 10 per cent. denser than the liquids.

This is a subject on which we ought to have better information, and I do think that physicists are at fault for not having given information regarding it to geologists. Speaking, on the other hand, in the capacity in which you do me the honour to place me, we, the geologists, are at fault for not having demanded of the physicists experiments on this subject. This time last year I urged that some one should take up the subject. I now renew the suggestion. We have a great Government fund for scientific inquiry. Why do not physicists and geologists unite in an inquiry like this, and apply for assistance out of the Government grant fund? They should then repeat Bischoff's experiments and

[1] Since this Address was delivered, some important experiments have been carried out at the request of Dr. Henry Muirhead, by Mr. Joseph Whitley, of Leeds. His experiments were made on iron, copper, and brass, and on whinstone and granite, and the general result, seemingly (but I believe not at all surely), indicated is that these substances are *less* dense in the solid than in the liquid state at the melting temperature.

test his results, but on a far larger scale, and extend the investigation to many materials besides those which he used.

In the meantime, the only experimental evidence we have, being Bischoff's, is to the effect that the melted rock is of less specific gravity than the solid rock, and that the solid rock would therefore sink down when frozen. Without insisting on this as being proved, I shall just call your attention to what would be the probable history of the solidification of a molten globe if this were the case. As soon as the surface began to freeze, and to freeze in sufficient quantity not to be floated up by mere superficial solidified foam, the mass of rock would fall down towards the centre. More would then solidify at the surface. This also would fall down, and the same thing would go on again and again. Gradually a sort of honeycombed solid would be formed. By and by a skeleton or frame-work through the whole would mount up to an extent sufficient to build up piers, as it were, to the surface, and the spaces between these piers when close enough would, in the continued freezing of the lava, be bridged across by solid rock thick enough in proportion to breadth not to break down and sink. There would again be breaking away of the piers and upheavals of the liquid material below; but by degrees the honey-combed mass would become nearly like a solid throughout with comparatively small interstices of liquid lava. This is in part the view of Hopkins. On account of arguments drawn from the phenomena of precession and nutation he was led to regard the earth as rigid on the whole; and looking to the existing evidence as to geological phenomena he was led to hold a view of the probable history of the earth agreeing with that which I have sketched, at all events so far as it would lead to the supposed present honeycombed condition.

Hopkins considered other very important questions, and among these was the question of the influence on their condition as to solidity and fluidity, of the great pressure to which the interior parts of the globe are subjected. He made experiments as to the influence of pressure on the melting points of some fusible substances, though he was not able to extend them so far as to test the effect of pressure in altering the melting point of rock materials.

Without making direct experiments, however, as to the influence of pressure on the melting point of rocks, we are able to infer by thermo-dynamic reasoning what must be the effect of pressure on the temperature of fusion, if we know whether rocks expand or contract in the act of solidifying. My brother, Professor James Thomson, has pointed out that it is a necessary consequence of thermo-dynamic laws that the temperature of the melting-points of substances which expand in solidifying should be raised by the application of pressure, while the opposite would be the case with substances which contract in becoming solid. Water, for instance, expands when it freezes: and he calculated the amount by which the freezing-point of water ought to be lowered by the application of an additional atmosphere of pressure. He found it to be about $\frac{1}{40}$ of a degree Fahrenheit ; and his theoretical result was afterwards verified by experiments made in the incipient Physical Laboratory in the old buildings of the University of Glasgow. Now, if the result announced by Bischoff, that the rocks he examined contract during solidification, be correct, it would follow as a consequence of thermodynamic laws that these substances melt at a higher temperature under high pressure than that at which they melt under low pressure. The conclusion to be drawn respecting the internal condition of the earth is, that we are not to infer liquidity of the interior, even if we should find evidence of a much higher internal temperature than that which would melt the rocks under ordinary pressure.

The view which I wish to put before you just now, however, is of a very different nature. I mean to deny altogether the intensely high temperature which Hopkins accepted, and the liquefying effect of which he endeavoured to obviate by introducing considerations regarding the solidity of the rocks, in consequence of intense pressure even at very high temperature. It is too generally supposed that the rate of increase of underground

temperature as we proceed inwards is uniform, or
nearly so, and that as we penetrate mile after
mile we should still find the temperature increasing
with equable rapidity. Thus, if we take metrical
measurement and the centigrade scale of thermometry, we find an increase in temperature as
we go downward of 1° C. per 30 metres of descent.
This, if we suppose the rate of augmentation of
temperature uniform would give 1000 degrees at
30,000 metres, and 3333 degrees at a depth of 100
kilometres. One hundred kilometres is about 60
statute miles ; and thus at 60 miles below the
surface we should have a temperature of about
3000 degrees centigrade. This is a temperature
far higher than would be required to melt any of
the ordinary rock materials at ordinary pressure.
But what I want to ask now is, "What reason
have we for supposing that, as we proceed farther
and farther in, the temperature does go on increasing at this rate ?" A very fallacious answer
has been given: "No falling off in the rate of
" increase of underground temperature has been
" found at the greatest depth to which observa-

" tion has reached, and therefore none is to be
" expected at greater depths, at all events so far
" as 50 or 100 miles down "

This question was very carefully discussed[1] in
§ A of the British Association at its Glasgow meeting in 1876; and I showed that instead of there
being reason to expect in an equable rate of increase
of temperature downwards in the greater depths, we
have on the contrary very strong reason to believe
that at a depth of 30 kilometres the rate of increase
downwards is considerably less than it is at the surface, and that at the greatest depths the temperature is not high enough for fusion of the material.

The state of the case then is, that we have no
reason from anything that has been observed, or
that could have been observed, at depths to which
we have already penetrated, for concluding that the
rate of increase of underground temperature, observed at small depths beneath the surface, continues
uniformly as we proceed inwards ; and that any
argument for intensely high temperature within,
based on the assumption that the rate of increase

[1] See pp. 240 244 of present volume.

does exist undiminished at considerable depths, falls to the ground.

The accompanying diagrams, drawn to scale (*Natural Philosophy*, Thomson and Tait, Vol. I., Appendix D) show the *rate of augmentation*, and the *augmentation* of underground temperature as we proceed downwards, on the assumption that the true rate of augmentation of temperature, after the superficial layer affected by summer heat and winter cold has been passed, is 1° Fah. per 51 feet at the beginning, and that the diminution in rate of augmentation of temperature downwards through the upper crust is 10 per cent. per 30 kilometres, or 100,000 feet. Fig. 1 shows the varying rate of augmentation of temperature downwards to a depth of 180 kilometres. Distances measured in the direction OX are depths below the surface, the unit length being 400,000 feet. Thus the position of the point P' in the diagram represents the state as to rate of augmentation of temperature at about ·7 of 400,000 feet, or 57,000 feet, below the surface. Ordinates measured outward on the diagram from the line OX represent rate of augmentation of

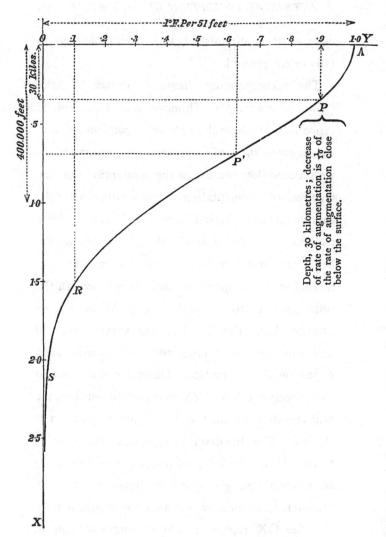

Fig. 1. Curve showing Rate of Augmentation of Temperature downwards.

underground temperature, the length OA being 1°
F. per 51 feet. Thus at P' the rate of augmentation
is between six- and seven-tenths of 1° F. per 51 feet,
and you must notice how rapidly the falling away
takes place. At the point R, which is about 600,000
feet below the surface, the rate of augmentation is
only one-tenth of a degree per 51 feet, and at
S, about 800,000 feet below the surface, it is
sensibly zero. Beyond this point underground
temperature does not sensibly increase.

Fig. 2 shows the excess of temperature above the
surface temperature at various depths, assuming the
rate of augmentation of temperature used in Fig. 1.
The depths are measured along OX, and the temperature at each depth is measured outward from
the line OX to the curve. The length OA represents 7900° F   Thus you will see that the excess
of temperature at P', 600,000 feet below the surface, is not so much as three times as high as at
P, which is, 100,000 feet below the surface, while at
S, less than 1,000,000 feet below the surface, a
maximum temperature of 7000° F. is reached.

Now, I do not presume to fix within any limits,

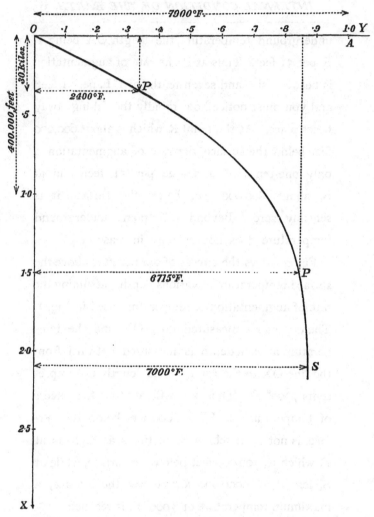

FIG. 2. Curve showing Excess of Temperature above Surface Temperature, assuming the rate of augmentation of temperature indicated in Curve I.

even of rough approximation, what the greatest
temperature reached in going downward may be.
It may be less than 4000° F., or 5000°. It may
possibly be as much as 8000° or 10,000°; but there
is a vast difference between five or ten thousand
degrees and the one hundred thousand or million
degrees, which we frequently find in geological
disquisitions, and against which we do not find
sufficient argument, or any argument at all, so far
as I am aware, in any regular geological teachings,
or published papers or books.

It appears, then, from the considerations that I
have brought before you, that we must give up the
argument derived from the phenomena of underground temperature as to the internal fluidity of
the earth. They show that the earth's interior is
at a high temperature, but not at any temperature
so high as to make general rigidity of the earth's
interior impossible or improbable. This being
the state of the case, we are forced to look to arguments drawn from other sources to decide the
question as to solidity or fluidity,—arguments such
as are supplied to us from the phenomena of

precession and nutation of the earth and from the phenomena of the tides. These arguments have been discussed in detail in a paper on the Rigidity of the Earth,[1] and in Thomson and Tait's *Natural Philosophy*, Vol. I.; and more recently in my Address to the Mathematical Section of the British Association at its meeting in Glasgow.[2] The arguments derived from the phenomena of precession and nutation present considerable difficulties, and indeed do not afford us at the present time a decisive answer. The phenomena of the tides, however, lead us to no uncertain conclusion. Suppose the earth to consist of a thin shell or crust enclosing, or floating on, a vast interior of molten matter. The liquid interior would tend to yield freely to the tide-generating influence of the sun and moon. The consequence would be that the exterior crust would be acted on by forces which, unless it were of preternaturally rigid material (I shall give you numbers directly), it would be unable

[1] "On the Rigidity of the Earth." W. Thomson, *Trans. R.S* May, 1862.
[2] See pp. 240, 244 above, in the present volume.

to resist. The crust would then be subject to upheavals and depressions taking place in time with
the revolutions of the sun and moon. If the crust
yielded *perfectly*, there would be no tides of the sea,
no rising and falling relatively to the land, at all.
The water would go up and down with the land,
and there would be no relative movement; and in
proportion as the crust is less or more rigid the
tides would be more or less diminished in magnitude. Now we cannot consider the earth to be
absolutely rigid and unyielding. No material that
we know of is so. But I find from calculation [1] that
were the earth *as a whole* not more rigid than a
similar globe of steel the relative rise and fall of the
water in the tides would be only $\frac{2}{5}$ of that which it
would be were the rigidity perfect ; while, if the
rigidity were no greater than that of a globe of
glass, the relative rise and fall would be only $\frac{2}{3}$ of
that on a pefectly rigid globe.

"Imperfect as the comparison between theory
"and observation as to the actual height of the
"tides has been hitherto, it is scarcely possible to

Thomson and Tait, *Natural Philosophy*, Vol. I., § 842.

" believe that the height is only two-fifths of what
" it would be if, as has been universally assumed in
" tidal theories, the earth was perfectly rigid.   It
" seems, therefore, nearly certain, with no other
" evidence than is afforded by the tides, that the
" tidal effective rigidity of the earth must be greater
" than that of glass."   This is the result taking the
earth as a globe uniformly rigid throughout.   That
a crust fifty or a hundred miles thick could possess
such preternatural rigidity, as to give to the mass,
part solid and part liquid, a rigidity, as a whole,
equal to that of glass or steel is incredible ; and we
are forced to the conclusion that the earth is not a
mere thin shell filled with fluid, but is on the whole
or in great part solid.

POLAR ICE-CAPS AND THEIR INFLUENCE IN CHANGING SEA LEVELS.

[*Being Paper read before the Geological Society of Glasgow, February* 16, 1888.]

THE subject I have to speak about this evening is not exactly geological. I may say that the immediate proposal to lecture on such a subject is to be found in an extract which I shall read to you from Dr. Croll's book on *Climate and Time.* In chaps. xxiii., xxiv., of this volume Mr. Croll deals with the physical causes of the submergence and emergence of land during the glacial epoch, and he has given some very curious, while at the same time mathematically correct, explanations of the effects due to a certain assumed displacement of ice from one hemisphere to the other. After

loyally calling attention, in his opening words, to
the fact of his having been anticipated by M.
Adhémar, (in a work *Révolutions de la Mer,*) in the
suggestion of heaped-up ice being a probable cause
of the submergence and emergence of land, Mr.
Croll proceeds to investigate the probable effect
of an ice-cap of a given description. In this connection Mr. Croll refers to an article on the
subject published by him in the *Reader* for
January 13, 1866, and the extract which I will
now read to you from this volume, *Climate and
Time* (pp. 372-374), consists of a note written
by myself, at Mr. Croll's request, in regard to the
objection brought forward in that article :

" Mr. Croll's estimate of the influence of a cap of
" ice on the sea level is very remarkable in its rela-
" tion to Laplace's celebrated analysis, as being
" founded on that law of thickness which leads to
" expressions involving only the first term of the
" series of ' Laplace's functions,' or ' spherical har-
" monics.' The equation of the level surface, as
" altered by any given transference of solid matter,
" is expressed by equating the altered potential

" function to a constant. This function, when ex-
" panded in series of spherical harmonics, has for
" its first term the potential due to the whole mass
" supposed collected at its altered centre of gravity.
" Hence a spherical surface round the altered centre
" of gravity is the *first* approximation in Laplace's
" method of solution for the altered level surface.
" Mr. Croll has with admirable tact chosen, of all
" the arbitrary suppositions that may be made
" foundations for rough estimates of the change of
" sea level due to variations in the polar ice-caps,
" *the* one which reduces to zero all terms after the
" first in the harmonic series, and renders that first
" approximation (which expresses the *essence* of the
" result) undisturbed by terms irrelevant to the
" great physical question.

" Mr. Croll, in the preceding paper, has alluded
" with remarkable clearness to the effect of the
" change in the distribution of the water in in-
" creasing, by its own attraction, the deviation of
" the level surface above that which is due to the
" *given* change in the distribution of solid matter.
" The remark he makes, that it is round the centre

" of gravity of the altered solid and altered liquid
" that the altering liquid surface adjusts itself
" expresses the essence of Laplace's celebrated
" demonstration of the stability of the ocean, and
" suggests the proper elementary solution of the
" problem to find the true alteration of sea-level
" produced by a given alteration of the solid.   As
" an assumption leading to a simple calculation,
" let us suppose the solid earth to rise out of the
" water in a vast number of small flat-topped
" islands, each bounded by a perpendicular cliff,
" and let the proportion of water area to the whole
" be equal in all quarters.   Let all of these islands
" in one hemisphere be covered with ice, of thick-
" ness according to the law assumed by Mr. Croll
" —that is varying in simple proportion to the sine
" of the latitude.   Let this ice be removed from the
" first hemisphere and similarly distributed over the
" islands of the second.   By working out according
" to Mr. Croll's directions, it is easily found that
" the change of sea-level which this will produce
" will consist in a sinking in the first hemisphere
" and rising in the second, through heights varying

" according to the same law (that is, simple propor-
" tionality to sines of latitudes), and amounting at
" each pole to

$$\frac{(1-\omega)it}{1-\omega w},$$

" when t denotes the thickness of the ice-cap at the
" pole, i the ratio of the density of ice, and w that
" of sea-water to the earth's mean density ; and ω
" the ratio of the area of ocean to the whole
" surface.

" Thus, for instance, if we suppose $\omega = 2/3$, and
" $t = 6,000$ feet, and take $1/6$ and $1/5\frac{1}{2}$ as the
" densities of ice and water respectively, we find
" for the rise of sea-level at one pole, and depression
" at the other,

$$\frac{\frac{1}{3} \times \frac{1}{6} \times 6000}{1 - \frac{2}{3} \times \frac{1}{5\frac{1}{2}}}$$

" or approximately 320 feet.

" I shall now proceed to consider roughly what
" is the probable extent of submergence which,
" during the glacial epoch, may have resulted from
" the displacement of the earth's centre of gravity

v

"by means of the transference of the polar ice from "one hemisphere to the other."

I wish you to notice particularly the last sentence of that note, for it is to that my attention has been called by your secretary. I was quite unaware that by this statement I had placed myself under any obligation, but it so turns out. A friend suggests that the quotation marks as supplied to that last sentence be here deleted, and the passage given as Mr. Croll's own. Mr. Croll may well have adopted it, for after quoting my note he at once proceeds to carry out the intention which I expressed in the concluding sentence. For myself I can only say that I am now trying to fulfil my obligation, though I feel I can throw but little additional light on the subject. Nor do I need to do so. it is quite unnecessary for me to carry out the intention, since Mr. Croll has done it with all the means that occurred to him as bearing on former estimates, in regard to this very important and difficult subject.

With regard to the effect on sea-level Mr. Croll's principle, as set forth in pp. 368-369 of his book, is thoroughly correct, and shows the remarkable power he possessed of grasping the subject, and dealing with it by a simple geometric construction which led to the same result as Laplace's mathematical analysis. For the stability of the ocean it is necessary that the specific gravity of water be less than the specific gravity of the solid, and it *is* less as we know. The mean density of the earth is about 5½—or more exactly 5·6—times the specific gravity of water. This statement favours Laplace's theory as to the requisite for stability, but it is curious that Laplace did not notice the simple view that if the solid part of the earth had a specific gravity less than that of water it would tend to float and leave the water on either side of it. For example, take a globe of liquid of any size (Fig. 1) and let us suppose a small spherical portion at A to become solid. The configuration of solid and liquid would remain stable provided the solidified portion remained of the same density as the liquid. If the solidified portion acquired a density greater or less than that of the liquid the configuration as shown in the figure would become unstable, and the small sphere A would take the position A' or A", in either of

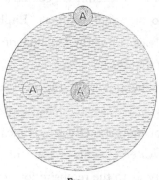

FIG. 1.

which two cases the configuration would be again stable.

Look now at Fig. 2 and suppose the outer circle to represent the section of a globe 8,000 miles in diameter, and let the inner circle represent the section of an enclosed globe 2 miles less in diameter. Suppose the outer envelope to be water, and the inner globe to be solid matter, of density equal to or greater than that of water, then the configuration of Fig. 2 would evidently be stable. On the other hand, if we suppose the

inner globe to be solid but of density less than
that of water, say equal to that of cork or wood,
then the water would no longer form an envelope
enclosing the solid, but would run together, and

Fig. 2.

the configuration of solid and liquid would be
represented by a section such as Fig. 3.

Mr. Croll goes on to inquire what would be
the probable effect, upon the level of the ocean, of
changing the centre of gravity of the earth, if we
suppose a quantity of ice (a polar ice-cap) to be
transferred to one pole from the other. I shall just
say one word as to the attraction of the polar ice-
cap.

We have here (Fig. 4) a globe. Suppose that somehow or other a portion of ice was placed on the Antarctic continent, what would be the result? It would be that the polar ice-cap would attract the water so that the water which stood at a certain height before that transference was

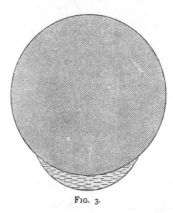

Fig. 3.

made, would be drawn up to a higher level all round the Antarctic continent by the attraction of this mass of ice. The calculation for the result is merely a piece of mathematical book-keeping with which I need not trouble you.

There is just one other point which belongs to further explanation of Laplace's theory. That first

theory of stability is really so clear that it is a
wonder Laplace did not see it straight away. But
he was so intensely mathematical that he often did
not see the simplicity of the results which he
attained by complicated mathematical analysis.
Suppose you shift a quantity of ice from one place
to another on the globe; and suppose, instead of

FIG. 4.

sea-water of its actual density, we had ideal water
of a twentieth part of the density of sea-water, then
the attraction of this solid mass of ice upon the
water would be calculated simply by the attraction
of the ice upon the ideal water. The figure would
be such that the surface of the ideal water would
be everywhere perpendicular to the surface of the

globe. But when we deal with real sea-water we have a piece of very delicate and nice mathematical book-keeping which reminds one of certain rules in Compound Interest. The ice attracts the fluid it displaces; but the displaced fluid itself attracts the remaining fluid and so contributes to the resultant attractive force. First calculate the amount of the attraction on the water due to the ice-cap alone; then calculate the increase of the attractive force due to the displaced water; then calculate this increase, to the second degree of approximation, and so on; and thus you get at the result. There have been different calculations founded on largely varied assumptions for data. Mr. H. D. Heath and the Rev. O. Fisher made calculations which differed somewhat widely from mine as well as from Croll's, but I believe them all to be consistent. My result, 380 feet, seemed to be immensely smaller than the others. In point of fact, Croll omitted to notice that mine referred to an ice-cap gathering on an ideal set of islands, and that I supposed the whole land to be distributed uniformly, and the ice to be placed upon the top of these islands. I took the actual proportion of the area of land to water, roughly estimated, as being one-third land and two-thirds water. To bring my result into comparison with the others quoted, you must therefore treble it, because I took only one-third area as covered with ice. Three times 380 = 1140, which is my number on a certain supposition; while Heath's is 650 on a supposition not quite the same as mine Pratt's estimates is still greater—something like 2,000 feet.

Now, if I could say anything to throw light upon the real question of extensions of ice in the southern or northern polar regions, and the effect of such extensions upon the sea level and upon the climate in past times, I should feel my attempt was certainly not insignificant. But I cannot even look upon such an attempt ; I can merely point out certain fallacies and set certain limits to former suppositions. We cannot have an ice-cap on the Antarctic continent 12½ miles thick, as Mr. Croll has calculated. I can bring substantial evidence against this. But Mr. Croll's argument does not at all stand upon that number; he is satisfied with a small fraction of it, 3,000 or 5,000, or 12,000 feet, instead of 65,000 feet.   He is satisfied with an ice-cap of a comparatively moderate thickness as a sufficient cause for some most important fluctuations of sea level which geological history proves to have taken place.   It seems to me that Croll is here meeting my case, and we may find the most probable explanation of some of our familiar changes of sea level—familiar even to people who are not geologists—in Croll's supposition of shiftings of ice either on the Antarctic or the Arctic hemisphere.

In the first place I shall ask you to imagine that the Antarctic continent for some unknown cause had at one time a distribution of ice over it thicker by 1,000 feet than at another time.   To be more accurate I would say 1,200 so as to correspond with the area equivalent to 1,000 feet of water. Mr. Murray has made a very careful estimate of the area of the Antarctic continent, which shows that the area is about one-fortieth of the area of the whole earth.   Croll makes it more ; but Mr. Murray has given us the more recent and more probable estimate. Now I shall merely ask you to think of a great ice-cap melted off the Antarctic continent, an ice-cap 1,200 feet thick equivalent to 1,000 feet depth of water. Imagine this mass of ice melting and flowing into the ocean , it would just raise the level of the ocean by one-fortieth of a thousand feet, a quarter of a hundred, or 25 feet. Our latest change of sea level here on the Firth of Clyde was only 10 feet. We do not know exactly the date, but it is quite certain that it was not very many thousand years ago. The water-level in the Firth of Clyde was then 10 feet higher than it is now, and that change of level would, on the theory I have stated, involve the melting of only about 400 feet of ice from the southern continent, which would raise the water 10 feet all over the world. When the theory of gravitation is taken into account in the manner I have indicated, the water thus brought to one of the poles and con-verted into an ice-cap, or the water that flows away from the melting ice-cap leaving a deficiency of solid water, does by its own gravitation always exaggerate the effect. Before we go into any consideration whatever regarding the possibilities of thickness of Antarctic ice, I think we may say that if it is 500 or 600 or 2,000 feet thick at one time, it may be at another time more or less, and if so we should have corresponding changes of level all over the world. We have had such perfectly feasible cases put before us by Taylor, Heath, and others as to the thickness of ice at either Pole, and in respect to the level of the sea.

Before taking up the question of how thick the ice may be at either Pole, think a little of the shape of the earth and ocean around the North Pole. There is the North Pole (see Fig. 5, giving an outline chart). Here is Greenland and Iceland. You enter into the Arctic Ocean west of Iceland ; then you come on to Spitzbergen, and further north to Nova Zembla, or again away to the east. This (the Arctic) ocean is merely a landlocked sea. The Behring Strait is only fifty miles wide and fifty fathoms deep, so that you may look upon the Arctic Ocean as practically stopped here. America is an island separated from the north-east of Asia by Behring Strait ; and Europe, Asia, and Africa

constitute another island, though Africa may be
considered a separate island because of the Suez
Canal. The Arctic Ocean connects with the North

FIG. 5.

Atlantic chiefly by the aperture between Norway
and Greenland, with the large island of Iceland—
one-third of the way across from Greenland and

two-thirds from Scotland. If a barrier were to be formed, as by raising slightly the bed of the ocean between the north of Scotland, Faroe Island, Iceland, and Greenland, the Arctic Ocean would become a lake. Such a barrier indeed already exists in a partial degree ; we see from this chart that there is a zone of comparatively shallow water all the way across from Norway to the Faroe Isles, Iceland and Greenland (see 500-fathom depth lines in Fig. 5) There is another opening by Davis Strait, Baffin's Bay, and Smith's Sound, and Captain Markham's expedition went up that narrow neck. If it also were closed the Arctic Ocean would practically become an inland lake.

Now what about the ice here ? In the first place so far as we know, the North Pole is under water. We know of no land north of Franz Josef Land. The farthest north that has yet been reached was by Captain Markham, who went 28 miles from the shore north of Smith's Sound, and reached the latitude of 83° 20′ 26″ N. On boring the ice it was discovered to be only 64 inches thick, and they found water—76 fathoms—beneath.

They went over floating ice, and only succeeded in getting over something like 30 miles ; for floating ice is exceedingly rough and hummocky in character, making the passage difficult for travellers. All we know then about the North Pole is that it is probably floating ice. There is also very strong, if not absolute evidence to show that there is great freedom for currents to flow under the ice across the polar region. Here is a piece of wood (specimen exhibited) from the banks of a Siberian river carried down in a floe of ice. That ice-floe with the wood was found in latitude 76° 30′ N., and longitude 40° W The pine-tree structure of the wood implies that it grew in a country far north. It came, as I said, embedded in an iceberg, and there is no possibility of that except by its being carried, by one of the great Siberian rivers which flow into the Arctic Ocean from the land of Siberia, acoss the North Pole into the latitude where it was found. There may be islands round the North Pole, but we know nothing of such. The Arctic Ocean, so far as known, has no very great island in the middle of it. This (second specimen shown) is a piece of another tree which was found fixed in an iceberg that floated across the North Pole, and there is the mud which was found on it. The wood is evidently a piece of a fir tree, with branches chiefly on the one side.

I want now to consider the physical properties of ice, so as to learn what limits, if any, these may give to a possible thickness of floating ice or of an icecap. Can ice stand in the form of any iceberg 1,500 feet thick? Icebergs have been said to stand at a height of 600 or 800 feet: but evidence is wanting to justify such estimates. The highest iceberg recorded is 700 feet, but there is a doubt whether it was even so high as that. What then would be the physical condition of a mass of ice at the North Pole? Think of the ice with snow falling on it at the rate of 3 feet annually. (I always reckon in feet of water, so that if I speak of 3 feet of snowfall, I mean sufficient to form 3 feet of water when melted.)

In considering this question I must first call your attention to Forbes's celebrated theory of the viscosity of ice, and his viscous theory of glaciers.

Forbes has demonstrated it by experiment, having studied the glaciers moving down the Swiss valleys and compared their motion with that of pitch. And now I am going to show you that motion in shoemaker's wax. Here is a piece (first model), you see what it has been doing since last Monday when it was placed a roughly rounded lump on the board : it is now flattened out like a pancake. Then here is another piece of shoemaker's wax that was placed, on the 2nd December, 1886, at the bottom of this jar with 10 bullets on the top and 14 corks placed below. If you look at the bottom of the jar you will see signs of bullets at the bottom, and you all see that the corks are making way through the wax and will be floating on the surface by and by. I may inspect the corks a little by carving into one of them. Probably in a fortnight that one will have taken advantage of this aperture I have just made. Thus, you see, in little more than a year the bullets have all disappeared beneath the wax, and the corks are coming to the surface.

But there is another experiment going on in this same jar. Three small cylinders of wood were each weighted by a piece of lead attached to one end, so that each mass—wood and lead—just floated on water. On the 4th February, 1887, these three cylinders were placed on the surface of the shoe-maker's wax, one with the weighted end down, another with the weighted end up, and the third on its side; and there you see what has been going on during the year. The first cylinder has sunk steadily into the wax; the second has toppled over and is sinking to a vertical position with the weighted end down; the third has attained a vertical position and is nearly as deep in the wax as the first.

I would call your attention to another illustration of the viscosity of this seeming solid. There is a piece of shoemaker's wax which was levelled down with heat and then allowed to become cold. Two months ago, on the 16th December, 1887, we placed *on* its surface the 16 pieces of wood—cylinders and cones—which you see are very slowly sinking into the wax. Here again I have pieces of wood shaped, and some of them with lead attached, exactly as are those on the wax. See the positions they take, and how quickly they do it, when I throw them on this fluid substance—water. Look at this cone in water. That cone (in wax), now so much inclined, was upright on 16th December, and I shall not say where it will be in the course of another month. *There* is a long cylinder in water —you observe it falls down on its side. Then here is a shorter cylinder in water—it falls into an inclined position. Here again we have a very short cylinder which would be clearly unstable with its edge down. There we have a cylinder with lead on it—it is stable. It is a counter-part of that which floats in water with the lead up, and would also float with the lead down, but it was placed in its present position. Now, there you have floating bodies which in a second of time, in water, do that which those similar bodies *floating* in wax are to be gradually doing in the course of the next few years. So that it is only a question of time between water and the shoemaker's wax.

Look at that globe of shoemaker's wax floating in water. Just watch it; if you could watch it with me for a month you would see what would happen.

But to economize your time look at this second one, which was placed in water last Tuesday exactly as you see the first—a globe. It is now like a pancake. It has flattened itself on the surface of the water to a round cake, of which the section is shown approximately by Fig. 6.[1] The water was

FIG. 6.

slightly heated, as it would take two or three years for the globe of wax to get as flat as that in cold water. We have hurried it up this evening by hotter water, but I am now going to leave it to itself.

Then here is a model—a wooden board having its edge shaped to represent the various forms of shore line, and covered with a layer of wax—illustrating the Arctic ice on the viscous theory. This was prepared last Monday, and has been somewhat hurried up by heat. You see the wax spreading itself on the plane, and tumbling over the edges.

[1] The curve is accurate for a floating *ice*-cap, and was drawn from the equation $y = b \left(1 - \dfrac{x^2}{a^2}\right)^{\frac{1}{4}}$ with b (height) = 2 inches, and a (diameter) = 20 inches.

Here is another model, which was only started to-day, placed in water so that the board is just at the water level. You see we have the continental features complete with a long slope down to the sea on one side, a deep gorge, a precipice, and yet another slope down into the sea ; and these variations of boundary are sufficient to show the various effects due to the shape of the land under a plastic body : everywhere ice flowing seawards breaking off when sea-borne, and making ice-bergs.

That shoemaker's wax is a substance which we know and see to be plastic ; here is a substance—ice—which we might not know to be plastic, but which Forbes proved to move as a plastic body in the Swiss glaciers. My brother's theory of the plasticity of ice, in virtue of melting by pressure and regelation, is admirably illustrated by an experiment of Mr. J. T. Bottomley. Here is a mass of clear ice having a piece of copper wire hung upon it, with a weight attached of 56 lbs. This wire will not go through the ice in less than an hour, but we see it already sinking into the ice—at the corners it is about half an inch in, and if we have time to look at it again we shall see the wire thoroughly embedded in the ice. This practically illustrates my brother's theory of the plasticity of ice—according to which one side of the wire presses into the ice, and in pressing cools and melts it. The ice relieved from pressure and the water coming together around the wire give rise to freezing above the wire. In general if pressure not equal in all directions be applied to ice, we have a solid which if it melts will relieve itself from that stress. A solid so circumstanced relieves itself from stress by inter-molecular melting ; and takes up a form free from stress by re-freezing.

In an investigation brought before the Royal Society of London last May by Dr. Main, it was stated that experiments made in the Engadine during February, 1887, with the temperature many degrees below freezing point, upon slabs of solid ice, showed that these yielded regularly as any other viscous body yields. Main's investigation gave perfectly definite results, though differing considerably with the temperature. He found that a bar of ice can be elongated. Take a bar of sealing wax ; hang a weight to it, and it will yield. So a bar of cold ice several degrees below freezing point—in an atmosphere at least 8 or 10 degrees below freezing point —was found to elongate by one third per cent. per 24 hours ; and the stress was 2 kilos. per square centimetre. So also look at the shoemaker's wax filling this model, which illustrates the principal features in the path of say a Swiss glacier : not warm, but cold as it is here now, it will keep yielding constantly from year to year ; working its way down this precipice, through this gorge, down this next precipice, out through this hole, and ultimately falling over the edges. Although it is brittle, it goes on day after day, month after month, yielding. I am not aware if my audience knows, what everybody knew many years ago, that a sealed letter left a long time seal down, flattens the seal with its own weight. We were always warned in those days not to seal letters for the tropics. Thus if you give it time enough, you will find the wax (alluding to the former model glacier) yielding, yielding, and yielding, according to the same laws, precisely as this mass of wax has yielded, aided by heat, in ten minutes—only with years instead of minutes. It would have taken months, I believe, instead of minutes, for an ordinary ball of shoemaker's wax to flatten in this way, if it had not been hurried by heat. I have made a calculation of this, but I will not trouble you with the somewhat intricate dynamics of the matter.

Turn again to this model showing the flattened ball : you may imagine this semi-fluid wax pulled in all round, held in as it might be by the sides of a containing box. Well it would then just take its level as water would. Now imagine the box placed in water and the sides taken away : the effect would be the same as if the wax were drawn out all round. There is no limit to the extent to which it would flatten itself, provided you did not keep adding material. But if you keep adding material, you arrive at a certain definite thickness. Suppose there to be a quantity of ice covering a large area, and all of uniform thickness. How is this uniformity to be preserved ? A great island covered with ice, 5 ft. thick and 10 miles across, with snow falling on it : at what rate must the snow fall upon it to keep

its thickness uniform ?   Taking the calculation of
viscosity of ice from Main's experiment, I find that
ice 10 metres thick would require 33 centimetres
per annum to fall upon it, to keep it of equal thickness.   A metre is 39·3708 inches, or say 40 inches.
The rate at which snow must fall upon ice 10
metres thick so as to keep its thickness uniform,
would be the equivalent to 33 centimetres of rain
per annum.   Double the thickness, or treble it, and
you would require an increased snowfall to pre-
serve it uniform—double thickness requiring a
quadruple fall, and so on.   Thus if you assume
a snowfall four times the amount stated, you would
only get double thickness.   So that if the thickness of the ice depended solely on its viscosity, you
may say that, between reasonable limits as to
amount of snowfall, the floating ice on the Arctic
Sea could not be thicker than from 10 to 20
metres ; that is to say, 10 to 20 yards = 30 to 60
feet.   This is about the thickness that it can have,
if there is nothing whatever carrying it away.

In a wholly enclosed Arctic Ocean such an ice
sheet would go on extending till it came to the

shores, and then it would go on getting thicker and thicker, till it would make an Arctic ice-cap like the Antarctic ice-cap. But the Arctic Ocean is not wholly enclosed by land. In the present condition of the Arctic Ocean there is an enormous abstraction of ice, by the Gulf Stream flowing in here (Fig. 5), and shooting away past Norway, and even carrying trees through the region of the North Pole. You may depend upon it that under the surface the ice is being simply washed away, and so kept down to 20, 30, or 60 inches in thickness. As I have said, Captain Markham's expedition found it only 64 inches thick within 399 miles of the North Pole, and we have no reason to believe that it is anything thicker, or very much thicker, at the very Pole. The free circulation through the Arctic basin, of water under the ice, is certainly what keeps down the thickness of the Arctic floating ice just now.

Suppose the circulation were stopped by a barrier running across from Norway by Iceland to Greenland, the result would be that the circulation of the water from the rest of the ocean into the

Arctic Sea would be stopped and the ice would spread out. The whole ocean would be cooled down to freezing point, and then the ice would thicken, the average temperature being far below freezing point. The mean annual temperature within the Arctic circle is 10 degrees (Fahrenheit). Mr. Buchan has kindly prepared for me an estimate of the temperature, which I shall read.

" Mean annual temperature within Arctic circle, as a whole, is 10° F.   Pole of cold appears to lie in about lat. 84° N., long. 150° W., around which an area of over 6,000 square miles has a temperature of − 5ᵘ F.   Two smaller centres of low temperature—one over Siberia about lat. 72° N., long. 123° E., and the other in North America, lat. 73° N., long. 105ᵘ W., over Melville Sound.   Each of these centres has a temperature slightly below 0° F.

" In January the centre of the main low temperature area is in almost the same position as shown on the annual map.   The lowest temperature shown is − 35° F., and this iso-thermal includes an area of nearly 2,000,000 square miles.

The subsidiary centre over Melville Sound also shows a temperature of − 35° F., and is in the same position as on the annual map. That over Siberia is farther to south and east, being now over Werchojausk, lat. 67° N., long. 134° E., and is of great intensity, the mean January temperature of Werchojausk, taken over four years, being −60° F.

" In July the lowest observed mean in the polar area is 35° F. No separate areas of low temperature appear, the coldest region to the north of Asia being in the Kara Sea, east of Nova Zembla. Low temperatures, 30° F. to 35° F., persist over Melville Sound, probably on account of the southerly direction of the ice-drift."

Under this condition it is perfectly clear that if the circulation of water from the rest of the ocean into the Arctic Sea were stopped, that sea would get filled up with solid ice, which would get thicker and thicker until we should have a prodigious ice-cap in the northern polar regions. We don't know if there was ever such a thing ; but I hope geologists will find it out, and I see no reason why their ingenuity, and their skilful and laborious scrutiny of palæo-historic monuments, should not discover it.

Let us consider now for a little the state

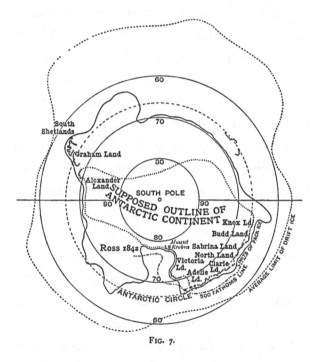

FIG. 7.

of matters at the South Pole. Just imagine (Fig. 7) this continuous black line to represent the boundary of the ice. It has been intimated that at certain parts the Antarctic ice-cap terminates in precipices of 170 to 200 feet in height. Mr. Croll's calculation is invalid beyond a certain limit, because it has been made on the supposition that the slope is uniform from the shore line inwards. He estimates that at the South Pole the thickness of the ice-sheet would be 12 miles. I say the slope cannot be uniform, and any reasoning, dependent on the assumption of uniform slope, must be fallacious.

Let us look at the realities of this Antarctic Continent in the light of what we know regarding the viscosity of ice. Through the kindness of Mr. Murray of the *Challenger*, I am enabled to place before you this splendid map (from which an outline sketch is given in Fig. 7). You see here Victoria Land, explored by Sir James Ross. We have here high mountains running up to a height of 8,000 feet, and a volcano, Mount Erebus, 12,000 feet high—also discovered by Ross. There is an ice-barrier running 200 miles towards the west, which is everywhere about 170 or 200 feet high above the sea. Here it is in our *model* Antarctic Continent already referred to. We break a piece off and thus we send an iceberg away. *There* is an ideal bay and gorge; a great high line of precipice, and ice slipping down into the water, and eventually breaking off: with ice below water nine times more than ice above it. There [showing a sketch of an iceberg] is an ice precipice 170 feet above, and below 1,530 feet; and here is the *Challenger* in a snow storm just after she has broken her jib-boom on the iceberg. The soundings taken here are described as giving 264 fathoms—just deep enough to float out one of these huge icebergs—at least they won't need to crawl along the bottom of the sea, if the neighbourhood of the shore there is anything like the neighbourhood of the shore elsewhere. In some places you get a depth of 250 fathoms very close to the shore, but this sounding may be about 100 miles from the shore. In ordinary water the ice will float out, when it gets into water deep enough to break off. Now let us take Mr. Murray's estimate of snowfall, which is worked out in an elaborate manner, founded upon the observations of expert German observers at South Shetland, just within the Antarctic circle. This would be something like four feet of snow and rainfall. (I am sorry I must give you results not always in metres or centimetres: easy things made difficult is the result of the English system of weights and measures.) I find, then, that with the viscosity taken from Main's observation, the thickness of the ice sheet at the South Pole would be 5,300 metres, that is about 18,000 feet. It is a very considerable thickness—about three miles—but only a quarter of Croll's estimate, which is much too high. The ice could not possibly stand on the Antarctic continent at a height of twelve miles.

Questions as to the possible effect of this Antarctic ice-sheet, or of changes in respect to its thickness, would be much too serious for me to enter upon just now. There are just two or three points I would like to consider. The amount of viscosity, and the probable slope of the Antarctic Continent, seemed to show that the ice cannot be very much thicker than 2,000 or 3,000 feet, and not nearly so much as 5,000 metres, as I have estimated

—though I don't say that so much is impossible
at the South Pole, but only that it seems improbable. It is, however, barely possible that the
absolutely different view taken by Mr. Croll may
be true. It is also possible that the land may
slope up to the South Pole, so that there may be
no ice at all upon it. We really do want to know
something about the South Pole. A memorial was
recently, as you will remember, sent to Government
asking them to assist in equipping an Antarctic
expedition. The request has not been granted,
but I think we may expect that Government will
yet see their way to it. We are encouraged all the
more, because we are assured it would be much
easier to drive a hansom up the hill of the South
Pole than up the hill to our University. The hill
of the South Polar ice-surface is, in all probability, (?) not more than a quarter of a degree, so
that one should easily drive up it! Ross believed
that if he could have found winter quarters there he
could have walked over the Pole. If you once get
ships up the coast there (pointing to the chart)
where Ross was, and get them comfortably frozen

A A 2

in, it might be found that there would not be much difficulty in arriving at the South Pole.

Among the first questions to be ascertained is, Whether there is bare rock or ice at the Pole, and what may be the effect of underground heat there? It is quite possible that the solid rock may slope up to a mountain, and not be covered with snow at all. I think it is covered with snow; but there may also be bare crags, as in regions where snow cannot lie, and by boring into them something may be learned of the underground temperature. If we bore down a thousand metres and find an increase of temperature at the rate of one degree centigrade per 27 metres, we may infer that it has been the same for many hundreds of years: if we find that the increase is smaller, we may be pretty sure that there has been a gain of ice; but if there is an increase of temperature of much more than that, we may infer an opposite state of things—that there has been melting, and that we are in a period subsequent to the melting. But take this case. Suppose the state of affairs to have been steady for about 5,000 or 10,000 years, or some little time like that, with an ice-cap 2,000 feet thick—and it is not at all improbable that that is the present condition—and suppose that under some of Croll's supposed astronomical changes snow began to fall twice as fast as now, and that it went on snowing for 2,000 years, the result would be that there would be a gain at the rate of, in round numbers, 2 feet per annum for 100 years—twice as fast as at present for 1,000 years. Set off against that the sliding down, and you get a very complicated problem. The terrestrial temperature would go for nothing. The thermal conductivity may be expressed thus :—How much heat would be conducted per annum through the 27 metres of rock ? You can make a calculation, remembering that there are $31\frac{1}{2}$ million seconds in a year. The temperature increases by one degree centigrade per 2700 centimetres, and the thermal conductivity of average rock (in gramme-water thermal units per square centimetre, per 1° per centimetre of rate of variation of temperature), is ·005. Thus we have the calculation :

$$\frac{31\frac{1}{2} \times 10^6}{2700} \times ·005.$$

The result is about 6 gramme-water-centigrade units of heat per annum. Now it would take about 79 such units to melt a centimetre of ice per annum. So that is all that underground heat would do; in other words, it would go for nothing in respect to retarding the increase of the ice-cap. This shows that if we were to have this change of temperature and this double snowfall for several hundred years, there would be a very sensible addition to the quantity of ice, and a very sensible depression on the water elsewhere. But if the Antarctic ice-cap were to be greatly increased it would lower the water, diminish the circulation, and tend to cool the Arctic Ocean. Therefore, from this considera-tion alone, we should expect glaciation in the northern hemisphere simultaneously with an aug-mentation of the Antarctic ice-cap.

But by far the most potent influence for altering the climate in any part of the world is oceanic circulation. The sea is the great carrier of heavy goods. Our atmosphere of air, with its pressure of fifteen pounds per square inch, corresponds to

33 feet of sea. A column of sea-water 33 feet long (or 10 metres, or 5½ fathoms) and square inch area, weighs 15 lbs. The atmosphere, mass for mass, is just equivalent to 5½ fathoms of sea, but 33 feet depth is a mere fraction of our sea-water. And again, a quantity of water is of much greater capacity for heat than the same quantity of air, the thermal capacity of air being very much less than that of water. Briefly then we may say that in the transport of heat over the whole solid globe the air goes for nothing. The sea is the great carrier. The air, of course, has an enormous indirect effect in the shape of gales, moderate winds, or trade winds, and in moving the water. It moves the surface of the water just as it moves the ships over the water. So that wind, indirectly helps the sea as a heat carrier. The wind is the distributer, the sea is the carrier—the great carrier—in the transport of heat

ON THE RATE OF A CLOCK OR CHRONOMETER AS INFLUENCED BY THE MODE OF SUSPENSION

[*Being a Paper read before the Institution of Engineers in Scotland, February* 27, 1867.]

IT is well known that the rate of a chronometer, a clock, or a watch may be altered by altering its mode of support. On land, clocks ought to be fixed in as solid a manner as possible, so as to prevent vibration, either by their own action or from extraneous causes, from being communicated to the supports of the pendulum. Even the best astronomical clocks hitherto made are not well arranged in this respect.

A marine chronometer or watch exhibits in a very striking manner the effects of varying the mode of support. A watch which keeps very good

time when carried in the pocket, or laid on a soft
pillow, will go at a different rate if laid on a marble
slab, or on a hard board. These variations of rate
are not due to any imperfections of the balance-
wheel or mechanism of the watch or chronometer,
but arise from reaction due to the motion of the
moving parts. A well-balanced watch will go
equally well whether supported in a vertical or
horizontal plane ; and a well-made watch will, I
believe, not be subject to uncertainty of above a
quarter of a second per day, if carried about in the
pocket all day and put under the pillow at night.
This I can testify from experience of a good
pocket-watch which I have tried now for nearly
two years ; indeed, a good pocket-watch, if well
treated, is comparable in its performances with
the best marine chronometer.

I was very much struck some time ago by a
remark made to me by Mr. Archibald Smith, of
Jordan Hill, regarding a demi-chronometer, with
detached lever and compensated balance, presented
to him by the Admiralty for the voluntary assistance he had given them in working out methods

for adjusting the compasses of iron ships. Mr. Smith found that this watch was going well, until one day he observed it had gained fifteen seconds, the reason of which he could not explain until he had recollected that instead of its having been put under the pillow as usual, it had been hung up in a suspended watch-case.

The question now arises, What is the cause of these variations, and how on dynamic principles are they to be explained? The dynamics of the subject are indeed very simple, and can be easily reduced to a well-known general problem.

A simple pendulum when it vibrates through a very small arc, vibrates according to the law of simple harmonic motion. Take a spiral spring, with a heavy weight hanging by it, stretch it a little and let it go, and it vibrates according to the same law. The vibrations of a tuning-fork, or any other instrument giving a similar musical sound, are also according to the law of simple harmonic motion. A case of roughly approximately simple harmonic motion we have when the piston moves to and fro in a cylinder, the head of the piston-rod being guided by a cross-head and slides, and the crank and fly-wheel making one revolution for every backward and forward movement of the piston. The balance-wheel of a watch, vibrating to and fro through a certain angle, performs very approximately a simple harmonic motion. The longer the hair-spring is, the more nearly it will approach to simple harmonic motion, and it will keep time the more accurately.

Now, against every change of motion of a body there is a certain reaction, and every motion to and fro of the balance-wheel of a watch or chronometer reacts upon the case of the watch or chronometer; and if the case is so suspended as to be free to vibrate, the motion of the balance-wheel will generate a vibration of the whole, so that we have two motions to consider—one, that of the balance-wheel inside the watch; the other, that of the whole watch except the balance-wheel. Upon the mode of suspension of the watch or chronometer will depend the nature of the vibration which it takes up and the resultant effect upon the rate. The rate is accelerated or retarded according as the vibration of the case is in the opposite direction to that of the balance-wheel, or in the same direction ; and the amount whether of acceleration or retardation, may be as much as a minute an hour, as I hope to demonstrate to you practically.

If a watch or chronometer be allowed extreme freedom to move, it has always a faster rate than when the case is held quite fixed. Mr. Archibald Smith has made experiments on this point upon a pocket-watch, with chronometer escapement and compensated balance, and found that the moment of inertia of the frame was 650 times that of the balance-wheel, from having observed that when hung horizontally by a long thread it had a gaining rate of some sixty-seven seconds in the day.

Observations made by Daniel Bernoulli on the sympathy of vibrations[1] manifested by the pans hanging from the two ends of a common balance, and the solution by Euler of the particular

---

[1] See a Paper (May, 1840), "On the Sympathy of Pendulums," by Mr. Archibald Smith, in Vol. II. of the *Cambridge Mathematical Journal.*

problem thus presented, seem to have originated the great dynamic problem of the vibrations of stable systems.

When a system of particles displaced from a position of equilibrium experiences in consequence forces in simple proportion to the displacements of its different parts, its motion may be thoroughly investigated by a generalisation of this problem of Bernoulli and Euler. The solution involves an algebraic equation of the same degree as the number of independent motions which may be given to the system. When the roots of this equation, which are necessarily all real, are all positive, the equilibrium of the system is stable. It is convenient to confine our attention to this case; but it is interesting and important to remark that all the statements we make in reference to it are applicable by a proper mathematical extension of the language, to cases of unstable equilibrium. Each of the roots of the algebraic equation used in other formulæ belonging to the solution, determines a particular proportion of different possible displacements, which, if made simultaneously, will give rise to *corresponding* forces of restitution according to the following condition. The system, starting from rest in its displaced configuration, will, under the influence of these forces, move so as to diminish the displacements of all its parts in the same proportion. Thus all the displacements will come to zero simultaneously; and therefore the system will move precisely through its configuration of equilibrium. There being no frictional or other resistance, it will oscillate—each displacement varying from maximum positive to maximum negative according to the simple harmonic law; the system passing, twice in each period, through its configuration of equilibrium, and being twice for an instant at rest in the configuration of extreme displacement on either side. This is called a fundamental mode of vibration. There are as many such fundamental modes as the system has of degrees of freedom to move (independent variables). Every possible motion of the system may be resolved in simple harmonic vibrations according to these fundamental modes; or the superposition of simple harmonic vibrations, according to the fundamental modes, will give any possible motion of the system. The arbitrary circumstances of displacement and projection by which any possible motion of the system may be instituted are producible by giving proper values to the energies and proper times to the epochs of maximum displacement of the component fundamental modes. The squares of the periodic times of the fundamental modes are the roots of the algebraic equation referred to above. In particular cases, some of these periods may be equal to one another; or all may be commensurable. In general, however, the periodic times of the fundamental modes are all different and incommensurable; and then none of the compound motions—that is to say, no motion except one or other of the fundamental modes—is periodic. The mathematics of the problem, including proofs of these results, will be found in the first volume (now on the point of appearing) of Thomson and Tait's *Elements of Natural Philosophy*.

The theory is not limited to systems presenting a finite number of independent variables, such as two in the cases we are about to consider more particularly, but is applicable to flexible or elastic bodies and fluids ; and to complex systems presenting a finite number of independent variables, on account of solid bodies or material particles, and infinite numbers of variables, due to flexible, elastic, or fluid matter, influenced by them. It includes, for example, the well-known dynamical theory of the vibrations of a stretched cord, of air in an organ pipe, or of water in an open basin of any shape. In the first two of these cases the periods are all sub-multiples of the gravest fundamental modes ; whence the explanation of the harmonics of musical cords and of wind instruments ; whence also the fact that a stretched cord struck or disturbed in any manner takes a perfectly periodic motion, and gives a true, although not a pure and simple, musical sound, with the peculiar character of the violin, pianoforte, or harp, depending on the way in which the vibration is excited. But the fundamental modes of vibration of an elastic solid—for instance a stiff metal bar, or a stiff spiral wire (as the "bell" of an American clock), a sheet of metal, or a common bell, are incommensurable. Hence these bodies cannot give any true musical sound other than a pure and simple harmonic note. A large sheet of metal, or a gong, or a drum, when struck, produces an infinite number of discordant notes sounding simultaneously (hence used for "music of demons" in lyric theatres!). But in the drum, the gravest of the fundamental notes predominates more decidedly, than does any one of the fundamental notes in the two other cases ; and thus a drum gives a nearer approach to a true musical sound than a sheet of brass or a gong.

An excellent illustration of the general theory is presented by the double pendulum— one pendulum hung from the weight of another. If we admit only vibrations in one plane, the system has two degrees of freedom to move. The determinant equation becomes a quadratic with two roots, necessarily unequal. The mathematics need not be given here ; but may be advantageously worked out as an exercise by the dynamical student. In the graver fundamental mode the two cords deviate always in the same direction from the vertical; the lower through a greater angle than the upper. In the quicker fundamental mode, the two deviate in opposite directions. The period of the graver fundamental mode is always longer than that of a simple pendulum, of length equal to that of the longer of the two cords; the period of the quicker fundamental mode is always shorter than that of the simple pendulum, equal in length to the shorter cord. If the upper mass is much greater than the one hung from it, and if the two strings be not approximately equal in length, the two fundamental periods differ but little from those of simple pendulums equal in length to the two cords respectively. The diagrams—Figs. 1 to 4— illustrate the circumstances in the cases; first, when the upper cord is considerably longer than the lower; and second, when the lower cord is considerably longer than the upper. In each case OA is the length of the the simple pendulum vibrating in the same period as that of the fundamental mode represented.

CASE I.

Figure 1 represents the first or graver fundamental mode ; the period of the upper pendulum

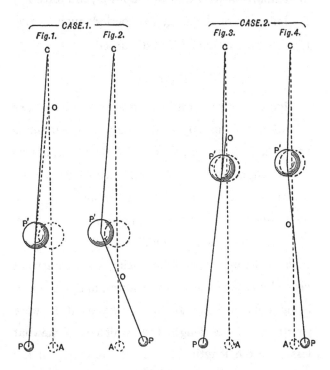

CP' being made somewhat graver by the influence of the lower, which, in the course of the vibration always exerts a force upon it *from* its middle position. Figure 2 represents the second or quicker fundamental mode ; the vibration of the upper pendulum being in this instance excessively small in comparison with that of the lower, and forced by the influence of the latter to a period much smaller than its own would be if undisturbed.

CASE II.

Figure 3 represents the graver mode. The vibration of the upper pendulum through but a very small arc in comparison with the lower, has its period augmented by the influence of the lower, which, in the course of the vibrations, exerts a force upon it always *from* its middle position. Figure 4 represents the quicker mode ; the vibrations of the upper pendulum being made somewhat faster by the influence of the lower, and the lower being influenced so as to vibrate as if it were shortened to the length OA, which is somewhat less than the length CP'. If P' consisted of the frame and work of a spring clock, and P' P were its pendulum, then, in Case I., the vibrations which would be maintained by the action of the escapement wheel would be that represented by Figure 2, and the clock would go faster than if its frame were perfectly fixed. In Case II., the vibrations maintained by the escapement would be those represented by Fig. 3, and the clock would go somewhat slower than its proper rate. Case I. could never occur in practice, but may be experimentally illustrated by hanging the works of a clock on a light stiff frame, movable round a horizontal axis. Case II., Fig. 3, with CP′ much shorter in proportion to P′ P than shown in the diagram, represents the actual circumstances of an ordinary pendulum clock, which, owing to want of perfect rigidity of the frame, must experience a little of the influence of the pendulum in the manner there illustrated, causing the rate of the clock to be somewhat slower than it would be if the support of the pendulum were absolutely fixed. The clock cases of the best astronomical clocks do not seem well adapted to give the steadiness necessary for good results ; and it is wonderful that their performances are not worse than they are found to be. The pendulum ought to be hung from a massive stone or metal support, attached to a stone pier, such as those used by astronomers for bearing their optical instruments. There can be little doubt but that the use of this simple precaution, and the making the pendulum many times heavier than hitherto, might render the performances of an astronomical clock, even with a Graham's dead-beat escapement, not merely two or three times better than those of a good watch carried about in the pocket, but ten or twenty times better, which we might well expect it to be in its immensely more advantageous circumstances. A good marine chronometer is probably little less accurate than the best astronomical clocks of the present day. It seems strange that such a very great improvement on Graham's dead-beat escapement as either the chronometer escapement or the detached lever constitutes, should not yet have been applied to the astronomical clock.

An interesting illustration of the influence of the different modes of suspension on the rate of a chronometer is had by suspending bifilarly a marine chronometer or a good pocket-watch. For

the suspension we may use stout silk threads
attached to the pivots of a marine chronometer
taken out of its gimbals, or attached to a pocket-
watch, at opposite points of its circumference,
by any convenient sling or by means of a cord
knotted round the watch like a parcel. The
suspending threads may be one, two, or three feet
long, and the upper ends may be attached to
adjustable points of support on a fixed horizontal
bar. We readily thus find that we can make it
go either fast or slow as we choose, by shifting
the points of support nearer to or farther from
the centre. When the points of support are
very near, the time of vibration of the chronometer as a whole if turned a little round its
vertical axis from the position in which it hangs
in equilibrium and let go, is much longer than
that of the balance-wheel. If the watch-case is
now steadied and left to itself, with the watch
going, the reaction of the balance-wheel, through
the spring, against the frame, gives rise to a
vibration, illustrated by Fig. 2, in which the
balance-wheel and the rest of the chronometer

vibrate round a vertical axis always in opposite directions. The effect of suspension in this instance is to make the watch go faster than when its case is held perfectly fixed, but this effect is smaller the nearer the upper points of support are. The circumstances of the extreme case when they are as close as possible are best realised by hanging the chronometer, as in Archibald Smith's experiment, by a long single cord, from a fixed point, by means of a sling or three short cords tied round the watch and so adjusted as to keep its face horizontal, thus giving the watch as a whole perfect freedom to move round a vertical axis. The permanent effect is then such, that the balance-wheel and the rest of the chronometer oscillate in opposite directions through ranges inversely as their moments of inertia. The period of this vibration is the same as that which the balance-wheel would have if the length of the hair-spring were diminished to the same proportion to its whole length that the moment of inertia of the chronometer (with the balance-wheel ideally free on its pivots) bears to the sum of this moment of inertia and the moment of inertia of the balance-wheel round its own axis. The period of vibration will be diminished according to square-root of this ratio. Thus, if the moment of inertia of the chronometer is 649 times that of the balance-wheel, the period will be $\sqrt{\frac{648}{650}}$, or about $\frac{1299}{1300}$ of the proper rate ; or the chronometer will gain one second in 1299, or about 67 seconds in the twenty-four hours. This was the result observed by Mr. Smith, from which he inferred the moment of inertia of the pocket chronometer referred to above.

If, on the other hand, the upper points of support are put very wide apart, the vibration maintained is of the same character as that illustrated in Fig. 3, and the watch goes slower than its proper rate. The farther apart the points of support are the less is this effect, as the circumstances approach more nearly to a perfect fixing of the frame.

If now, commencing with the upper points of support very close together, we gradually increase the distance between them, or, starting with them very wide apart, we gradually diminish the distance, a certain critical arrangement is approached from either direction, and the gaining rate in the former case, or the losing rate in the latter case, is augmented. This critical arrangement is such that the period of vibration of the suspended chronometer, when set to vibrate by an external disturbance, is approximately equal to the period of vibration of the balance-wheel. When the upper points of support are adjusted to produce it, and the chronometer, going, is left to itself, the action of the internal prime mover will bring the whole into a state of vibration, which may be either the first fundamental mode (balance-wheel and framework vibrating in the same direction), in which case the chronometer will have a losing rate, or the second fundamental mode (balance-wheel and frame vibrating in opposite directions), in which case the chronometer will have a gaining rate. The gain or loss may amount to as much as one second in sixty or eighty with an ordinary ship chronometer, taken off its gimbals, or a pocket detached lever watch. The amount of the effect will of course be much less for a marine chronometer, not removed from its gimbals, but suspended by cords attached to its outer case, on account of the great addition of moment of inertia due to the outer case. With a marine chronometer, or any watch having a chronometer escapement (Harrison's), or having a duplex escapement, the seconds hand jumps forward once, and one comparatively loud beat is heard, for each period of the balance-wheel; and thus it is easy to see whether the watch, when suspended, is vibrating according to the first fundamental mode (losing), or the second mode (gaining), by noticing in which direction the visible motion is at each beat of the escapement. With either of these kinds of escapement the experiments above described are liable to stop the watch when the upper points of support are adjusted for the critical arrangement. Thus, for instance, if the points of support have first been too close for the critical arrangement, and are gradually separated until the vibration of the frame becomes very large, a great gain of rate is produced; and if the distance is then a little farther increased, the watch will often stop : if then a slight impulse round the vertical axis is given to it to start it, it will commence vibrating according to the first fundamental mode, with a largely losing rate. The other corresponding result is obtained by commencing with the points of support too far asunder for the critical arrangement and bringing them gradually together.

Without exciting independent vibrations of the chronometer or watch as a whole, and counting them, it is easy to perceive whether the circumstances approach the critical condition, by applying the hand to steady the watch, and then observing the phenomena presented when it is left to itself. If the upper points of support are either much too wide apart or much too close together for the critical arrangement, the watch-case will not take any regular harmonic vibration, but will make a slight (perhaps scarcely perceptible) jump once every semi-period, or once every period of the balance-wheel, according to the character of the escapement. But if the upper points of support be set approximately to the critical arrangement, and the watch brought to rest and left to itself, it will be seen to commence vibrating through a gradually wider and wider arc until a maximum of vibration is attained. The amplitude of vibration will then diminish, but not to zero ; will increase to a second maximum smaller than the first ; will diminish to a second minimum not so small as the first minimum ; increase to a third maximum smaller than the second ; and so on, until, after several of these alternations, a sensibly steady state of vibration, very closely simple harmonic, is attained. How nearly the critical arrangement is approximated to, may be judged by counting the number of vibrations executed from starting to the first maximum, from the first maximum to the first minimum, and so on—the numbers being greater the nearer the adjustment is to the critical condition. I made these experiments first on board the *Great Eastern* during her last summer's cruise ; and it was curious, as an illustration of the general principle of the superposition of motions, to watch the various phenomena of vibration which the suspended watch presented, quite independently of the swinging due to the rolling of the ship.

When the top points of support are arranged precisely to the critical condition, I find that the watch will, of itself, take sometimes one mode of vibration, sometimes the other. But a very slight deviation in either direction from the critical arrangement suffices to do away with this indifference, and to insure that, when the watch is steady and left to itself, it will take up either always the gaining or always the losing mode of vibration. But even then it may be compelled to take up either mode by properly-timed touches with the finger, and it continues vibrating accordingly when left to itself. Thus, when the top points of support are adjusted, either precisely, or somewhat approximately, to the critical condition, the watch may be made to go either faster or slower than its proper rate, by applying the hand to cause it to take up either mode of vibration at pleasure, and then leaving it to itself. This last experiment ought not, however, to be pushed too far with a valuable chronometer, as the effort to make it take up a mode of vibration opposite to that which it takes up of itself, is liable to make the escapement-wheel trip and run round rapidly, escaping from the control of the balance-wheel and the escapement—this disturbance not being produced by any violent action of the hand, but by very gentle touches properly timed. No such derangement can, I believe, ever take place when the watch is hung in the manner described, and left at rest to take up whatever mode of vibrating it will, and no damage to the most delicate chronometer can result.

The knowledge of those facts may be of advantage—first, in pointing out a simple plan for setting a chronometer without touching the hands ; second, in showing how it ought to be supported, in regular use, so that it may go at a uniform rate and keep correct time. It is usual to place ship's chronometers on cushions, at sea, to guard against damage to the works, from tremors of the ship. If the cushion be moderately hard, the chronometer's rate does not (as I have found by trials on board the *Great Eastern*) differ sensibly from what it is when the chronometer is laid on a hard board, the instrument being of course always kept on its gimbals in its heavy outer case. If, however, the cushion is soft enough, the critical condition explained above may be reached or even passed ; and great variations of rate in either direction may be produced. Thus a certain degree of softness in the cushion may make the chronometer lose considerably ; and a still softer cushion may make it gain considerably ; and cushions softer yet would make the chronometer gain, although not so much. It is possible that an improvement in the practical performance of chronometers at sea may be attained by fixing the outer case of the instrument to a very heavily weighted base, this base being placed on an ordinary cushion.

At the conclusion of the paper, in answer to questions by the PRESIDENT, Mr. DAY, and Mr. DAVISON,

Sir WM. THOMSON said that the weight of the chronometer would influence the rate at which it would gain or lose by the oscillation; and it is therefore better for good time-keeping to have a massive watch-case than a light one. No doubt, the rate of an ordinary watch-chronometer is very much affected by railway travelling. His own pocket-watch gained from four to eight seconds in journeys to London and back. The railway carriage vibration affected as a prime mover the vibration of the balance-wheel, not merely as vibrations induced in the frame by the interior movement would do. If a chronometer case is well weighted, its performance will not be practically injured by the influence which has been described. If it were firmly attached to the middle of a two-feet-long plank, with heavy weights fixed on it near the ends, its rate would be sensibly the same as if its case were absolutely fixed, however this board is supported. To avoid damage from the tremors of the ship, this board should be placed on cushions, and strapped down, or lashed properly, for security.

If a watch be hung on a nail, it depends upon the dimensions of the watch and the time of the balance-wheel whether it will go faster or slower than its proper rate. If, when hung on a nail and set to swing, it vibrates more rapidly than the balance-wheel, then the effect of the hanging would be to induce a slower rate; but if when set to swing it vibrates slower than the balance-wheel, then when left to itself it will go faster than when the case of the watch is held quite fixed. A watch regulated to go correctly when hanging on a nail (according to a faulty practice, sometimes followed, he believed, in watchmakers' shops) cannot be expected to go at even approximately the same rate as when carried about in ordinary use.

ON A NEW ASTRONOMICAL CLOCK

[Being a Paper read before the Royal Society, June 10, 1869.]

IT seems strange that the dead-beat escapement should still hold its place in the astronomical clock, when its geometrical transformation, the cylinder escapement of the same inventor, Graham, only survives in Geneva watches of the cheaper class. For better portable time-keepers, it has been altered (through the rack-and-pinion movement) into the detached lever, which has proved much more accurate. If it is possible to make astronomical clocks go better than at present by merely giving them a better escapement, it seems almost certain that one on the same principle as the detached lever, or as the ship-chronometer escapement, would improve their time-keeping.

But the inaccuracies hitherto tolerated in astronomical clocks may be due more to the faultiness of the mercury compensation pendulum, and of the mode in which it is hung, and of the instability of the supporting clock-case or framework, than to imperfection of the escapement and the greatness of the arc of vibration which it requires ; therefore it would be wrong to expect confidently much improvement in the time-keeping merely from improvement of the escapement. I have therefore endeavoured, though perhaps not successfully, to improve both the compensation for change of temperature in the pendulum, and the mode of its support, in a clock which I have recently made with an escapement on a new principle, in which the simplicity of the dead-beat escapement of Graham is retained, while its great defect, the stopping of the whole train of wheels by pressure of a tooth upon a surface moving with the pendulum, is remedied.

Imagine the escapement-wheel of a common dead-beat clock to be mounted on a collar fitting easily upon a shaft, instead of being rigidly attached

to it. Let friction be properly applied between the
shaft and the collar, so that the wheel shall be
carried round by the shaft unless resisted by a force
exceeding some small definite amount, and let a
governor giving uniform motion be applied to the
train of wheel-work connected with this shaft, and
so adjusted that, when the escapement-wheel is
unresisted, it will move faster by a small percentage
than it ought to move when the clock is keeping
time properly. Now let the escapement-wheel,
thus mounted and carried round, act upon the
escapement, just as it does in the ordinary clock.
It will keep the pendulum vibrating, and will, just
as in the ordinary clock, be held back every time
it touches the escapement during the interval
required to set it right again from having gone too
fast during the preceding interval of motion. But
in the ordinary clock the interval of rest is considerable, generally greater than the interval of motion.
In the new clock it is equal to a small fraction of
the interval of motion: $\frac{1}{300}$ in the clock as now
working, but to be reduced probably to something
much smaller yet. The simplest appliance to

count the turns of this escapement-wheel (a worm, for instance, working upon a wheel with thirty teeth, carrying a hand round, which will correspond to the seconds' hand of the clock) completes the instrument ; but if desired, minute and hour-hands, though a superfluity in an astronomical clock, can easily be added.

In various trials which I have made since the year 1865, when this plan of escapement first occurred to me, I have used several different forms, all answering to the preceding description, although differing widely in their geometrical and mechanical characters. In all of them the escapement-wheel is reduced to a single tooth or arm, to diminish as much as possible the moment of inertia of the mass stopped by the pendulum. This arm revolves in the period of the pendulum (two seconds for a one second's pendulum), or some multiple of it. Thus the pendulum may execute one or more complete periods of vibration without being touched by the escapement.

I look forward to carrying the principle of the governed motion for the escapement-shaft much further than hitherto, and adjusting it to gain only about $\frac{1}{100}$ per cent. on the pendulum ; and then I shall probably arrange that each pallet of the escapement be touched only once a minute. The only other point of detail which I need mention at present is that the pallets have been, in all my trials, attached to the bottom of the pendulum, projecting below it, in order that satisfactory action with a very small arc of vibration (not more on each side than $\frac{1}{100}$ of the radius, or 1 centimetre, for the seconds' pendulum) may be secured.

My trials were rendered practically abortive from 1865 until a few months ago by the difficulty of obtaining a satisfactory governor for the uniform motion of the escapement-shaft ; this difficulty is quite overcome in the pendulum governor, which I now proceed to describe.

Imagine a pendulum with single-tooth escapement mounted on a long collar loose on the escapement-shaft just as described above—the shaft, however, being vertical in this case. A square-threaded screw is cut on the upper quarter of the length of the shaft, this being the part of it on which the collar works, and a pin fixed to the collar projects inwards to the furrow of the screw, so that, if the collar is turned relatively to the shaft, it will be carried up or down, as the nut of a screw, but with less friction than an ordinary nut. The main escapement-shaft just described is mounted vertically. Below the screw and long nut-collar, the escapement-shaft is surrounded by a tube which, by wheel-work, is carried round about five per cent. faster than the central shaft. This outer shaft, by means of friction produced by the pressure of proper springs, carries the nut collar round along with it, except when the escapement-tooth is stopped by either of the pallets attached to the pendulum. A stiff cross piece (like the head of a T), projecting each way from the top of the tubular shaft, carries, hanging down from it, the governing masses of a centrifugal friction governor. These masses are drawn towards the axis by springs, the inner ends of which are acted on by the nut-collar, so that the lower or the higher the latter is in its range, the springs pull the masses inwards with less or with more force. A fixed metal ring coaxial with the main shaft holds the governing masses in, when their centrifugal forces

exceed the forces of the springs ; and resists the
motion by forces of friction increasing approximately in simple proportion to the excess of the

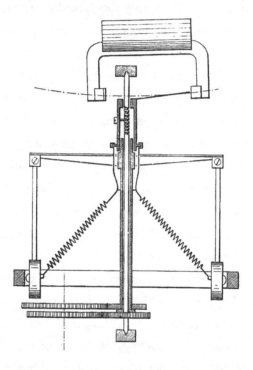

speed above that which just balances the forces of
the springs. As long as the escapement-tooth is
unresisted, the nut-collar is carried round with the
quicker motion of the outer tubular shaft, and so it

screws upwards, increasing the force of the springs.
Once every semiperiod of the pendulum it is held
back by either pallet, and the nut collar screws *down*
as much as it rose during the preceding interval of
freedom when the action is regular; and the
central or main escapement-shaft turns in the
same period as the tooth, being the period of the
pendulum. If, through increase or diminution of
the driving-power, or diminution or increase of the
coefficient of friction between the governing masses
and the ring on which they press, the shaft tends
to turn faster or slower, the nut collar works its
way down or up the screw, until the governor is
again regulated, and gives the same speed in the
altered circumstances. It is easy to arrange that
a large amount of regulating power shall be implied
in a single turn of the nut collar relatively to the
central shaft, and yet that the periodic application
and removal of about $\frac{1}{50}$ of this amount in the
half period of the pendulum shall cause but a *very
small* periodic variation in the speed. The latter
important condition is secured by the great moment
of inertia of the governing masses themselves round
the main shaft.

ON BEATS OF IMPERFECT HARMONIES

[Being Paper read before the Royal Society of Edinburgh, April 1st, 1878.]

ACCORDING to a usage which has been adopted from the German of Helmholtz by the best English scientific writers on sound, a sound is called a "simple tone,"[1] or without qualification a "tone," when the variation of pressure of the air in the neighbourhood of the ear, which is the immediate excitant of the sense, is according to a simple harmonic function of the time ; that is to say, when the deviation from the mean pressure of the air varies in simple proportion to the distance, from a fixed plane, of a

[1] The old musical usage, according to which the word tone denotes an interval (the major tone or minor tone, or the mean tone of the tempered scale), though it unfortunately clashes with this recent scientific use of the word tone, can scarcely be abandoned.

point moving uniformly in a circle. Considering the actual sensibility of the human ear to musical sounds, we must introduce farther as a practical restriction that the period of the variation of the pressure must be less than $\frac{1}{30}$ of a second, and greater than $\frac{1}{10000}$ or $\frac{1}{20000}$ of a second. The vibrations of the air produced by a simple harmonic vibrator are either simple harmonic, or are in circular or elliptic orbits, resulting from the composition of two simple harmonic motions ; and the consequent change of air-pressure in the neighbourhood of the ear follows the simple harmonic law, provided the maximum velocity of the vibrator and of the air in its neighbourhood be infinitely small in comparison with the velocity of sound. Hence the more nearly this condition is fulfilled the more exactly a simple tone is the sound heard ; but it is far from being fulfilled when the vibrator, though itself performing simple harmonic motion, has sharp edges round which the air is forced to rush with great velocity, or when, as in the case of free-reed organ pipes or the reeds of a harmonium, the vibrator is an elastic solid moving to and fro in a very narrow aperture. (In the case of a slapping reed, as of trumpet stops in an organ, the motion of the vibrator itself is not simple harmonic, and the sound is excessively rich in overtones, giving it its peculiarly rich or harsh character.)

A harmony is any sound of which the excitant change of air-pressure is strictly periodic, and is not a simple tone. According to Fourier's beautiful analysis[1] of periodic variations, to which the name of the harmonic analysis has been given, any periodically varying quantity may be regarded as the sum of quantities varying separately according to the simple harmonic law, in periods respectively equal to the main period, half the main period, a third of the main period, and so on. According to this analysis we see that the variation of air-pressure constituting a harmony may be regarded as the sum of variations constituting simple tones, one having its period equal to the

[1] Compare *Trans. R.S.E.*, April 30th, 1860 ; re-published in Vol. III. of my Mathematical and Physical Papers, " Reduction of Observations of Underground Temperature," where a short description of Fourier's analysis is to be found.

period of the harmony ; a second, half that of the harmony ; a third, one-third that of the harmony, and so on ; in other words, we may regard the harmony as compounded of these simple tones.

Practically, in musical language the term harmony is not applied when the tone of the main period predominates in the sensory impression, and in this case the sound is simply called a note ; its pitch is reckoned according to the main period ; and the effect of the other tones, now called overtones, which enter into its composition, are merely felt as giving it its character or quality of sound. Thus the name harmony is in musical practice restricted to cases in which there is either no tone of the main or fundamental period, or not enough to produce a predominating impression ; and a sound compounded of two, three, four, or more simple tones, having commensurable periods, is heard. In ordinary musical language a harmony is not regarded as having any one pitch, but is thought of as compounded of its known constituents. The true period of the harmony is, however, in every case the least common multiple of the period of its constituent tones. The number of times that the period of the harmony contains the period of any one of its constituent tones I call the harmonic number of that tone. This expression is only applicable to any particular tone when viewed as one constituent of a harmony. Following the usage of Lord Rayleigh and Professor Everett, I shall employ the word "frequency" to denote the number of periods per unit of time,—per second let us say generally in acoustical reckonings. Thus the "frequency" of a tone or of a harmony means the number of its periods per second. Similarly the frequency of any set of beats, according to the definitions and descriptions below, will mean the number of the beats per second, and in this application of the term it will designate sometimes a proper fraction, and sometimes a small whole number plus a proper fraction.

The quality of a harmony, when the periods of its several constituent tones are given, depends upon the amplitudes of the different constituents, and on the relation of their phases. Thus, for example, consider a harmony of two tones. They may be so related in phase that at one of the instants of maximum pressure of one of the constituents there is also maximum pressure of the other constituent. The same phase-relation, if the harmonic numbers of the constituent tones be both odd, will give also coincident minimums. But when one of the harmonic numbers is even and the other odd the phase-relation of coincident maximums will also be such that there is a coincidence of minimum pressure due to one tone with maximum pressure due to the other; and again there will be an opposite phase in which there will be coincidence of minimums, and in this opposite phase there will also be a coincidence of maximum and minimum. (To avoid circumlocutions a harmony of two odd numbers will be called an odd binary harmony; a harmony of even and odd numbers will be called an even binary harmony.) Thus we see that in an odd binary harmony there is a phase-relation of coincident maximums and coincident minimums, and again an opposite phase-relation of coincident maximum

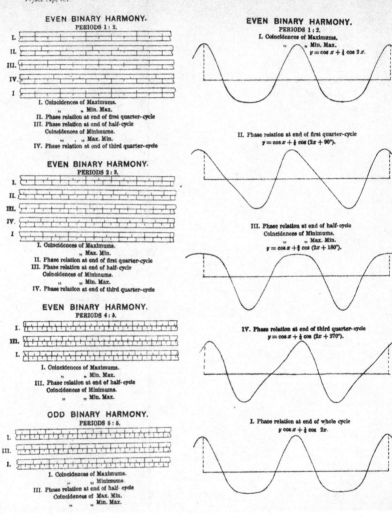

EVEN BINARY HARMONY.
PERIODS 1 : 2.

I.
II.
III.
IV.
I

I. Coincidences of Maximums.
„ „ Min. Max.
II. Phase relation at end of first quarter-cycle
III. Phase relation at end of half-cycle
Coincidences of Minimums.
„ „ „ Max. Min.
IV. Phase relation at end of third quarter-cycle

EVEN BINARY HARMONY.
PERIODS 2 : 3.

I.
II.
III.
IV.
I

I. Coincidences of Maximums.
„ Max. Min.
II. Phase relation at end of first quarter-cycle
III. Phase relation at end of half-cycle
Coincidences of Minimums.
„ „ Min. Max.
IV. Phase relation at end of third quarter-cycle

EVEN BINARY HARMONY.
PERIODS 4 : 5.

I.
III.
I.

I. Coincidences of Maximums.
„ „ Min. Max.
III. Phase relation at end of half-cycle
Coincidences of Minimums.
„ „ Min. Max.

ODD BINARY HARMONY.
PERIODS 3 : 5.

I.
III.
I.

I. Coincidences of Maximums.
„ „ Minimums.
III. Phase relation at end of half-cycle
Coincidences of Max. Min.
„ „ Min. Max.

EVEN BINARY HARMONY.
PERIODS 1 : 2.
I. Coincidences of Maximums.
„ „ Min. Max.
$y = \cos x + \frac{1}{4} \cos 2x.$

II. Phase relation at end of first quarter-cycle
$y = \cos x + \frac{1}{4} \cos (2x + 90°).$

III. Phase relation at end of half-cycle
Coincidences of Minimums.
„ „ Max. Min.
$y = \cos x + \frac{1}{4} \cos (2x + 180°).$

IV. Phase relation at end of third quarter-cycle
$y = \cos x + \frac{1}{4} \cos (2x + 270°).$

I. Phase relation at end of whole cycle
$y = \cos x + \frac{1}{4} \cos 2x.$

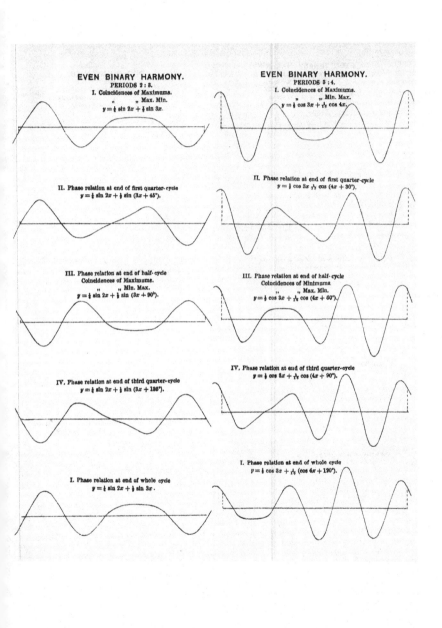

EVEN BINARY HARMONY.
PERIODS 2 : 3.
I. Coincidences of Maximums.
„ „ Max. Min.
$y = \frac{1}{2}\sin 2x + \frac{1}{3}\sin 3x$.

II. Phase relation at end of first quarter-cycle
$y = \frac{1}{2}\sin 2x + \frac{1}{3}\sin (3x + 45°)$,

III. Phase relation at end of half-cycle
Coincidences of Maximums.
„ „ Min. Max.
$y = \frac{1}{2}\sin 2x + \frac{1}{3}\sin (3x + 90°)$.

IV. Phase relation at end of third quarter-cycle
$y = \frac{1}{2}\sin 2x + \frac{1}{3}\sin (3x + 135°)$.

I. Phase relation at end of whole cycle
$y = \frac{1}{2}\sin 2x + \frac{1}{3}\sin 3x$.

EVEN BINARY HARMONY.
PERIODS 3 : 4.
I. Coincidences of Maximums.
„ „ Min. Max.
$y = \frac{1}{3}\cos 3x + \frac{1}{12}\cos 4x$,

II. Phase relation at end of first quarter-cycle
$y = \frac{1}{3}\cos 3x + \frac{1}{12}\cos (4x + 30°)$,

III. Phase relation at end of half-cycle
Coincidences of Minimums.
„ „ Max. Min.
$y = \frac{1}{3}\cos 3x + \frac{1}{12}\cos (4x + 60°)$.

IV. Phase relation at end of third quarter-cycle
$y = \frac{1}{3}\cos 3x + \frac{1}{12}\cos (4x + 90°)$.

I. Phase relation at end of whole cycle
$y = \frac{1}{3}\cos 3x + \frac{1}{12}(\cos 4x + 120°)$.

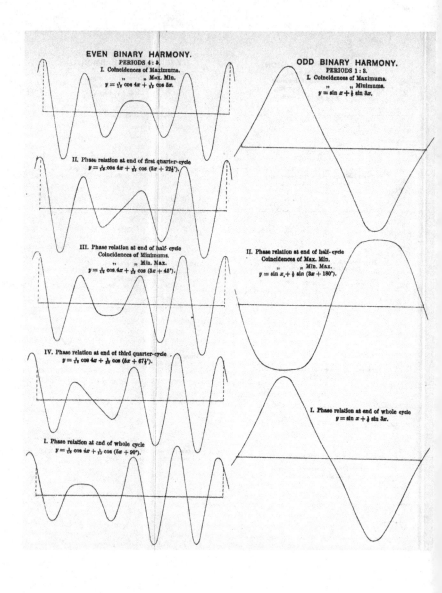

EVEN BINARY HARMONY.
PERIODS 4 : 5.
I. Coincidences of Maximums.
    "   "   Max. Min.
$y = \tfrac{1}{18}\cos 4x + \tfrac{1}{31}\cos 5x.$

II. Phase relation at end of first quarter-cycle
$y = \tfrac{1}{18}\cos 4x + \tfrac{1}{31}\cos (5x + 22\tfrac{1}{2}°).$

III. Phase relation at end of half-cycle
Coincidences of Minimums.
    "   "   Min. Max.
$y = \tfrac{1}{18}\cos 4x + \tfrac{1}{31}\cos (5x + 45°).$

IV. Phase relation at end of third quarter-cycle
$y = \tfrac{1}{18}\cos 4x + \tfrac{1}{31}\cos (5x + 67\tfrac{1}{2}°).$

I. Phase relation at end of whole cycle
$y = \tfrac{1}{18}\cos 4x + \tfrac{1}{31}\cos (5x + 90°).$

ODD BINARY HARMONY.
PERIODS 1 : 3.
I. Coincidences of Maximums.
    "   "   Minimums.
$y = \sin x + \tfrac{1}{3}\sin 3x.$

II. Phase relation at end of half-cycle
Coincidences of Max. Min.
    "   "   Min. Max.
$y = \sin x + \tfrac{1}{3}\sin (3x + 180°).$

I. Phase relation at end of whole cycle
$y = \sin x + \tfrac{1}{3}\sin 3x.$

ODD BINARY HARMONY.

PERIODS 3 : 5.

I. Coincidences of Maximums.
" " Minimums.

$y = \frac{1}{3} \cos 3x + \frac{1}{5} \cos 5x$

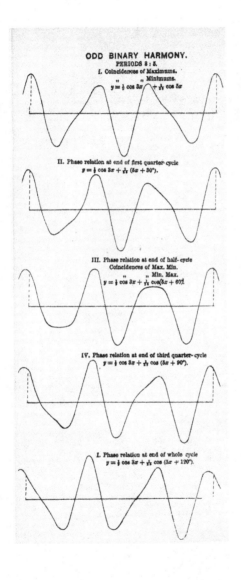

II. Phase relation at end of first quarter-cycle
$y = \frac{1}{3} \cos 3x + \frac{1}{5} (5x + 30°)$.

III. Phase relation at end of half-cycle
Coincidences of Max. Min.
" " Min. Max.
$y = \frac{1}{3} \cos 3x + \frac{1}{5} \cos(5x + 60°)$

IV. Phase relation at end of third quarter-cycle
$y = \frac{1}{3} \cos 3x + \frac{1}{5} \cos (5x + 90°)$.

I. Phase relation at end of whole cycle
$y = \frac{1}{3} \cos 3x + \frac{1}{5} \cos (5x + 120°)$.

minimum and minimum maximum. The former will be called the phase-relation of coincidences, the latter the phase-relation of oppositions. In an even binary harmony there is a phase-relation of coincident maximums and coincident maximum minimum; and again an opposite phase-relation of coincident maximum minimum and coincident minimums. The former will be called the phase-relation of coincident maximums, the latter the phase-relation of coincident minimums. The annexed diagrams illustrate and prove these assertions. The horizontal line in each may be either regarded as representing space at any one instant in the direction of propagation of the sound, or it may represent times at which successive phases of the motion are perceived by an ear in a fixed position. In the left-hand column of diagrams the long vertical cross-bar in each case denotes maximum of air-pressure; the short vertical cross-bar, minimum.

For lecture illustrations it is convenient to use long slips of wood with paper pasted on one side, and the short and long cross-bars marked upon it, and to support these slips of wood on a board with nails to guide them so that they may be placed in groups of two, three, or four, one over the other, and any of them moved in the direction of its length to illustrate the different phase-relations of a harmony.

Suppose now, one note of a perfect binary harmony to be very slightly sharpened or flattened : so slightly that during a large number of the periods of the perfect harmony, the phase-relation in the imperfect harmony experiences but little change. Let the two notes of the imperfect harmony be sustained long enough with perfect uniformity as to pitch and intensity :—the effect will be that of perfect harmony, modified by a slow change of its phase-relation through a cycle ; which in the case of an even binary harmony is from coincident maximums gradually to coincident minimums, and thence gradually round again to coincident minimums ; and in the case of an odd binary harmony is from oppositions to coincidences, and round to oppositions again ; and so on in cycles. In favourable circumstances, and with careful attention, a variation of the quality of the sound recurring periodically in these successive cycles is distinctly heard, even by an unpractised ear, unless the duration of the cycle be too long or too short to suit its sensibility. It is this variation which is called the "beat" on the imperfect harmony.

The period of the beat—that is to say, the duration of the cycle described above—is most easily found by taking the reciprocal of its frequency, calculated by the following rule :—*The frequency of the beat is equal to the error of frequency of one note multiplied by the harmonic number of the other.* When in a harmony of three or four notes all are perfect except one, the beats due to the imperfection of the false one are to be reckoned just as if the harmony were binary, according to the following rule :—

For the two or more notes which are in perfect harmony imagine one whose period is the period of their harmony. Take this as if it were one tone of an approximate binary harmony, the false note of the given harmony being the other. Example : Let the frequencies of the three notes be 257, 320, and 384 (an approximation to the harmony 256, 320, 384, or C, E, G.). The common period of the two last-mentioned is $\frac{1}{64}$ of a second, and we have to calculate the beats on two notes whose frequencies are 64 and 257. The harmonic numbers of the harmonies to which these notes approximate are 1 and 4, and the error in frequency of the higher note is 1 per second; hence the beats are at the rate of 1 per second. When there is error in two or more notes of a multiple harmony, two or more sets of beats in periods not commensurable with one another are heard; but the general effect is apt to be too confused to allow any one of the sets to be distinctly counted. On a multiple harmony with only one note false the beats are in general exceedingly distinct; more so in general than in binary harmonies.

Sometimes, as for distance in reckoning the beats in the imperfect harmonies of a tempered musical scale, it is convenient to regard the two notes of an imperfect harmony as in error from two notes of a perfect harmony differing but little from them; then the rule for calculating the frequency of the beats is to take the difference of the errors of the two notes, each multiplied into the harmonic number of the other. Thus, let n and n' be the harmonic numbers of the perfect harmony to which the given notes approximate, and let e and e' be the excesses of the vibrational frequencies of the two actual notes above two in perfect harmony nearly agreeing with them. The frequency of the beat of the actual notes is $n'e - ne$.

For example, take the following table of numbers of vibrations in a perfect diatonic scale, with 256 vibrations per second for C, and in the corresponding scale of equal temperament (founded on twelve equal semitones, in each of which the interval ratio is $2^{\frac{1}{12}}$) :—

Frequencies of True.		Frequencies of Tempered.		Error of Tempered.
C	256	256	C	0 0
D	288	287·35	D	− 0·65
E	320	322·54	E	+ 2·54
F	341·33	341·72	F	+ 0·39
G	384	383·57	G	− 0·43
A	426·67	430·54	A	+ 3·87
B	480	483·27	B	+ 3·26
C′	512	512	C′	0·0
D′	576	574·70	D′	− 1·30
E′	640	645·08	E′	+ 5·08

From these numbers we find the following table of beats :—

Perfect Harmonies.	Harmonic Numbers.	Imperfect Harmonies.	Qualities of Falseness of the Intervals.	Frequencies of Beats.
		Errors.		
G, C	3, 2	G − 0·43, C 0·0	} Too small	0·86
B, E	3, 2	B + 3·26, E + 2·54	} Too small	1·10
C', F	3, 2	C' 0·0, F + 0·39	} Too small	1·17
E', A	3, 2	E' + 5·08, A + 3·87	} Too small	1·45
F, C	4, 3	F + 0·39, C 0·0	} Too large	1·17
A, E	4, 3	A + 3·87, E + 2·54	} Too large	1·45
E', B	4, 3	E' + 5·08, B + 3·26	} Too large	2·20
E, C	5, 4	E + 2·54, C 0·0	} Too large	10·16
A, F	5, 4	A + 3·87, F + 0·39	} Too large	13·53
G, E	6, 5	G − 0·43, E + 2·54	} Too small	17·39
G, E	6, 5	G 0·0, E + 2·54	} Too small	15·24
C', A	6, 5	C' 0·0, A + 3·87	} Too small	23·22
A, C	5, 3	A + 3·87, C 0·0	} Too large	11·61
E', G	5, 3	E' + 5·08, G − 0·43	} Too large	17·39
E, G, C	5, 2	E + 2·54, G 0·0, C 0·0	} Too large	5·08

It is of course to be understood that the degree
of falseness is the same in all the tempered harmonies of the same name (or having the same
harmonic numbers); and that the different numbers shown for the frequencies of the beats are
(except for the case of the E with the untempered
G) in simple proportion to the vibrational frequencies of one or other of the constituent notes.
The slightness of the imperfectness in the tempered fifth (approximately 2 : 3) is indicated by
the slowness of the beats, not so much as one
per second on the C G. The imperfectness of the
fourth (approximately 3 : 4) is even less than that
of the tempered fifth, so that, notwithstanding the
greater harmonic numbers, the beats are scarcely
more rapid (1·17) for the C F than (86) for the
C G. But when we go to major and minor thirds
of the tempered scale, we take leave of mathematical harmony entirely. The beats on the
C E (ten per second) are too rapid to be counted,
and it is only in virtue of their not being perceived, or not being disagreeably perceived, that
the combination is agreeable. The same may be

said still more unqualifiedly of the minor thirds, the number of beats on $E\ G$ being more than seventeen per second. It does not seem easy to explain on any physical or physiological principles the decidedly agreeable effect produced on the ear by a succession of major and minor thirds of pianoforte notes. It is, no doubt, to the slowing of the beats by the superposition of a third note upon either of the binaries $C\ E$ or $E\ G$ in the ternary combination $C\ E\ G$, because of its comparatively close approximation to $\overset{\frown}{C\ G}\ E$ (for which the beats are only five per second), that the comparatively smooth harmoniousness of the common chord in the tempered scale is due.

It is not generally known how easily beats on approximations to other harmonies than unison are heard, even when the constituent notes are simple tones. Through the kindness of Professor M'Kendrick I have been allowed the means of testing them in very varied combinations, by aid of a series of excellent tuning-forks of Koenig's, each mounted on a wooden box resonator, after the manner of Marloye. For such experiments

Koenig's tuning-forks are much superior to Marloye's, because of the greater quantity of metal in
each fork, in virtue of which it gives a louder and
more enduring sound. The sound proceeding
from such a source is essentially a simple tone,
or very nearly so. I have tested that in every
case the number of beats counted is the smallest
that could be according to the preceding theory ;
for it is to be remarked that the theory only gives
the whole period of the phenomenon, but does not
answer the question—Does the ear perceive a
gradual variation of quality through the whole
period, or does it fail to distinguish the difference
of quality between two halves of the period, or
between three-thirds of it, and so on? My experiments demonstrate that in every case the ear
does distinguish the two halves of the period of
each beat. Thus, for example, in the beat on an
approximation to the harmony (1 : 2) in which the
variation of air-pressure on the ear is represented
by the preceding curves [1] for four instants of the
period noted, I find that the ear distinguishes the

[1] See page 401.

quality of the sound represented by the sharp-topped and flat-hollowed curve from that represented by the flat-topped and sharp-hollowed curve. In the one case the pressure of air close to the ear rises very suddenly to, and falls very suddenly from, its maximum, and (as in cases of tides in which there is a long hanging on low water) there is a comparatively slow variation of pressure for a few ten-thousandths of a second on each side of the instant of minimum pressure ; in the opposite phase-relation there is a slow change before and after the time of maximum pressure, and a rapid change before and after the time of minimum pressure. In the former case the difference of maximum from mean exceeds the difference of minimum from mean ; in the latter the difference of the minimum from mean exceeds the difference of maximum from mean. If the ear could not distinguish between two such sounds, but could distinguish between either of them and the sounds represented by the first and third quarter-cycle curves, the number of beats would be twice the error of frequency of the higher note.

But I find that the number of beats is quite distinctly equal to the error of frequency of the higher note. I have found that the beats on the 1 : 2 harmony are most easily perceived when the intensity of the higher note is comparatively faint, as would be the case if they were explained by Helmholtz's theory that they are the beats on the approximation to unison which there is between the higher note and the first overtone of the lower note. But the simple-harmonic character of the two constituent tones at the entrance of the ear precludes the acceptance of this theory unless extended, as it has actually been by its author, to the interior mechanism of the ear.

Whatever may be the physiological theory by which the beats are to be explained, it is an interesting fact that the ear does distinguish, as it were, between push and pull on the tympanum in the manner illustrated by the preceding curves, not only for the case of approximation to the harmony 1 : 2, but for every other even binary harmony. I have heard distinctly the beats on approximations to each of the harmonies 2 : 3, 3 : 4, 4 : 5,

5 : 6, 6 : 7, 7 : 8, 1 : 3, and 3 : 5. The two last
mentioned, though sometimes less easily heard
than the beats on most of the others, are unmis-
takably distinct ; and by counting the numbers of
them in ten seconds or in twenty seconds, I have
ascertained that they, as do all the others, fulfil
the condition of having the whole period of the
imperfection, and not any sub-multiple of it, for
their period of audible beat. They are interesting
as being cases of odd binary harmonies. Before
making the experiments, I thought it possible that
what is heard in the beat might not make
distinction between the configurations II. and IV.
(first quarter phase and third quarter phase) : but
a *revolving character* which I perceive in the beat
is to me certainly distinct enough to prove that
the ear does distinguish between these configurations, which are one of them the same as the other
taken in the reverse order of time.

In every instance except the octave, the beat on
the approximation to a binary harmony is less
distinct than the beat on an approximation to a
ternary or higher multiple harmony with only one

note false.   It is not because of the comparative
slowness of the beat on the multiple harmony ;
for by taking alternately beats with one note
slightly false in a binary harmony, and the same
note made more false in a ternary or multiple
harmony to such a degree as to give the same
number of beats, I have always found the beats in
the latter case much more prominent than in the
former.   Thus by taking first the perfect harmony
C E G (4, 5, 6), and the three binary harmonies
C G (2 : 3), C E (4 : 5), E G (5 : 6), and flattening
slightly any one of the three notes by screwing on
a small mass of brass to either or to each prong
of the tuning-fork producing it, it is easy after a
little practice to count the beats on each of the
binary harmonies.   Thus, for example (supposing
$E_{,}$ to designate a note of a slightly lower pitch than
E), after a little practice it is easy to count the
beats on C $E_{,}$ and on the $E_{,}$ G, and to verify that
their frequencies are, the first of them four times,
and the second of them six times, the error of
frequency of the $E_{,,}$ and then to verify that
the much louder beats on the ternary har-

mony C E, G, are of half the frequency of
the former, and of one-third of the frequency of
the latter, and to verify absolutely that they
are of twice the frequency of the error of E,.

If when the approximate harmony C E, is
being sounded, and the beats on it heard, the
faintest sound of G is produced by a very
gentle excitation of the fork by the bow,
instantly a loud beat at half speed is heard. The
phenomenon is rendered very striking by alternately touching the top of the G fork by the bow
so as to stop its vibrations, and then drawing the
bow very gently for a fraction of a second [1] along
one side to re-excite them. It is marvellous how
small an intensity of the sound G is required to
give a smooth unbroken loud beat in the double
period. I have found it difficult to excite the G
gently enough to give the gradual transition from,

[1] In every case, to obtain regular beats, each tuning-fork, after
being set in vibration by the bow, must be left to itself. The
sound is sensibly graver as long as the bow is applied to augment
or sustain the vibration than when the fork is left free. Thus, if
two tuning-forks nearly, but not quite, in unison, are alternately
acted on by the bow and left free, the beats are less rapid during
the time the bow is applied to the higher fork, and more rapid
when to the lower, than when both forks are vibrating freely.

let us say for example, four uniform beats per second, through the case of four beats per second with every alternate beat somewhat louder, to the case of only every second beat perceptible, or, in all, two beats per second ; but it can be done, and the result is an interesting and instructive illustration of the reduction from the quick beat of the binary harmony to half speed, or to one-third speed, or to one-fifth speed, as the case may be, by the introduction of a third note. In the several cases I have found that I can, by making the added note faint enough, produce a succession of beats of which every second, or every third, or every fifth, as the case may be, is louder than the others, and that, as the intensity of the added note is gradually increased, the fainter beats become imperceptible, and a regular unbroken slow beat is heard distinctly alone, always in the theoretical time of the whole imperfection of the harmony. I have verified this distinctly in the cases of 1, 2, 3 ; 2, 3, 4 ; 3, 4, 5 ; 4, 5, 6 (as stated above); 5, 6, 7 ; and 6, 7, 8. I have not succeeded in hearing the beats on the approximations to the harmonies 8 : 9 and 9 : 10.

But the slow beat on the 8, 9, 10 (with vibrational frequencies 256, 288, 320), with any one of the three notes slightly flattened, is very remarkable. The sound is like that of a wheel going round with decided roughness of motion in every part of its revolution, but much rougher in one part than another, with a loudly perceptible periodic return of the roughness in the theoretical period of the approximate harmony.

The beats on the harmony C E G (vibrational frequencies 256, 320, 384), with any one of the three notes slightly flattened, are very perceptible; untrained ears hear them instantly the first time without any education, and the beat is heard almost to the very end of the sound when three of Koenig's forks, one of them, the C, for example, being slightly flattened by a brass sliding piece screwed to it, are excited by the bow to well-chosen loudnesses, and are then left free. The sound dies beating, the beats being distinctly heard all through a large room as long as the faintest breath of the sound is perceptible. The smooth melodious periodic moaning of the beat is particularly beautiful when the beat is slow (at the rate, for instance, of one beat in two seconds or thereabouts), being, in fact, sometimes the very last sound heard when the intensities of the three notes chance at the end to be suitably proportioned.

ON THE

ORIGIN AND TRANSFORMATION OF MOTIVE POWER.

[*Being Friday Evening Lecture before the Royal Institution, February* 29, 1856; *from Proc. R. I. republished in Vol. II. of Mathematical and Physical Papers.*]

THE speaker commenced by referring to the term *work done*, as applied to the action of a force pressing against a body which yields, and, to the term *mechanical effect produced*, which may be either applied to a resisting force overcome, or to matter set in motion. Often the mechanical effect of work done consists in a combination of those two classes of effects. It was pointed out that a careful study of nature leads to no firmer conviction than that work cannot be done without producing an indestructible equivalent of mechanical effect. Various familiar instances of an apparent loss of mechanical effect, as in the friction, impact, cutting, or bending of solids, were alluded to, but especially that which is presented by a fluid in motion. Although in hammering solids, or in forcing solids to slide against one another, it may have been supposed that the alterations which the solids experience from such processes constitute effects mechanically equivalent to the work spent, no such explanation can be contemplated for the case of work spent in agitating a fluid. If water in a basin be stirred round and left revolving, after a few minutes it may be observed to have lost all sensible or otherwise discernible signs of motion. Yet it has not communicated motion to other matter round it ; and it appears as if it has retained no effect whatever from the state of motion in which it had been. It is not tolerable to suppose that its motion can have come to nothing ; and until fourteen years ago confession of ignorance and expectation of light was all that philosophy taught regarding the vast class of natural phenomena, of which the case alluded to is an example. Mayer, in 1842, and

E E 2

Joule, in 1843, asserted that heat is the equivalent obtained for work spent in agitating a fluid, and both gave good reasons in support of their assertion. Many observations have been cited to prove that heat is not generated by the friction of fluids : but that heat *is* generated by the friction of fluids has been established beyond all doubt by the powerful and refined tests applied by Joule in his experimental investigation of the subject.

An instrument was exhibited, by means of which the temperature of a small quantity of water, contained in a shallow circular case provided with vanes in its top and bottom, and violently agitated by a circular disc provided with similar vanes, and made to turn rapidly round, could easily be raised in temperature several degrees in a few minutes by the power of a man, and by means of which steam power applied to turn the disc had raised the temperature of the water by 30° in half an hour. The bearings of the shaft, to the end of which the disc was attached, were entirely external ; so that there was no friction of solids under the water, and no way of accounting for the heat developed, except by the friction in the fluid itself.

It was pointed out that the heat thus obtained is not *produced from a source*, but is *generated ;* and that what is called into existence by the work of a man's arm cannot be matter.

Davy's experiment, in which two pieces of ice were melted by rubbing them together in an atmosphere below the freezing point, was referred to as the first completed experimental demonstration of the immateriality of heat, although not so simple a demonstration as Joule's ; and although Davy himself gives only defective reasoning to establish the true conclusion which he draws from it. Rumford's inquiry concerning the "Source of the Heat which is excited by Friction" was referred to as only wanting an easy additional experiment—a comparison of the thermal effects of dissolving (in an acid for instance), or of burning, the powder obtained by rubbing together solids, with the thermal effects obtained by dissolving or burning an equal weight of the same substance or substances in one mass or in large fragments—to prove that the heat developed by the friction is not *produced from the solids* but is *called into existence between them.* An unfortunate use of the word "capacity for heat," which has been the occasion of much confusion ever since the discovery of latent heat, and has frequently obstructed the natural course of reasoning on thermal and thermo-dynamic phenomena, appears to have led both Rumford and Davy to give reasoning which no one could for a moment feel to be conclusive, and to have prevented each from giving a demonstration which would have established once and for ever the immateriality of heat.

Another case of apparent loss of work, well known to an audience in the Royal Institution— that in which a mass of copper is compelled to move in the neighbourhood of a magnet—was adduced ; and an experiment was made to demonstrate that in it also heat appears as an effect of the work which has been spent. A copper ball, about an inch in diameter, was forced to rotate rapidly between the poles of a powerful electro-magnet. After about a minute it was found by

a thermometer to have risen by 15° Fahr.   After
the rotation was continued for a few minutes more,
and again stopped, the ball was found to be so hot
that a piece of phosphorus applied to any point of
its surface immediately took fire.   It is clear that
in this experiment the electric currents, discovered
by Faraday to be induced in the copper in virtue
of its motion in the neighbourhood of the magnet,
generated the heat which became sensible.   Joule
first raised the question, Is any heat generated by
an induced electric current in the locality of the
inductive action ?   He not only made experiments
which established an affirmative answer to that
question, but he used the mode of generating heat
by mechanical work established by those experiments, as a way of finding the numerical relation
between units of heat and units of work, and so
first arrived at a determination of the mechanical
value of heat.   At the same time (1843) he gave
another determination founded on the friction of
fluids in motion ; and six years later he gave
the best determination yet obtained, according to
which it appears that 772 foot pounds of work

(that is 772 times the amount of work required to overcome a force equal to the weight of one pound through a space of one foot) is required to generate as much heat as will raise the temperature of a pound of water by one degree.

The reverse transformation of heat into mechanical work was next considered, and the working of a steam-engine was referred to as an illustration. An original model of Stirling's air-engine was shown in operation, developing motive power from heat supplied to it by a spirit lamp, by means of the alternate contractions and expansions of one mass of air. Thermo-electric currents, and common mechanical action produced by them, were referred to as illustrating another very distinct class of means by which the same transformation may be effected. It was pointed out that in each case, while heat is taken in by the material arrangement or machine, from the source of heat, heat is always given out in another locality, which is at a lower temperature than the locality at which heat is taken in. But it was remarked that the quantity of heat given out is not (as Carnot pointed out it would be if heat were a substance) the same as the quantity of heat taken in, but, as Joule insisted, is less than the quantity taken in by an amount mechanically equivalent to the motive power developed. The modification of Carnot's theory to adapt it to this truth was alluded to ; and the great distinction which it leads to between reversible and not reversible transformations of motive power was only mentioned.

To facilitate farther statements regarding transformations of motive power, certain terms, introduced to designate various forms under which it is manifested, were explained. Any piece of matter or any group of bodies, however connected, which either is in motion, or can get into motion without external assistance, has what is called mechanical energy. The energy of motion [1] may be called either " dynamical energy," or " actual energy." The energy of a material system at rest, in virtue of which it can get into motion, is called " potential energy," or, generally, motive power possessed

[1] Shortly after the date of this Lecture, I gave the name "kinetic energy," which is now in general use. It is substituted for "actual" and for "dynamical" in the remainder of the text. (K. Nov. 21, 1893).

among different pieces of matter, in virtue of their relative positions, has been called potential energy by Rankine. To show the use of these terms, and explain the ideas of a *store of energy*, and of conversions and transformations of energy, various illustrations were adduced. A stone at a height, or an elevated reservoir of water, has potential energy. If the stone be let fall, its potential energy is converted into kinetic energy during its descent, exists entirely as the energy of its own motion at the instant before it strikes, and is transformed into heat at the moment of coming to rest on the ground. If the water flow down by a gradual channel, its potential energy is gradually converted into heat by fluid friction, and the fluid becomes warmer by a degree Fahr. for every 772 feet of the descent. There is potential energy, and there is kinetic energy, between the earth and the sun. There is most potential energy and least kinetic energy in July, when they are at their greatest distance asunder, and when their relative motion is slowest. There is least potential energy and most kinetic energy in January, when they are at their least distance, and when their relative motion is most rapid. The gain of kinetic energy from one time to the other is equal to the loss of potential energy.

Potential energy of gravitation is possessed by every two pieces of matter at a distance from one another; but there is also potential energy in the mutual action of contiguous particles in a spring when bent, or in an elastic cord when stretched.

There is potential energy of electric force in any distribution of electricity, or among any group of electrified bodies. There is potential energy of magnetic force between the different parts of a steel magnet, or between different steel magnets, or between a magnet and a body of any substance of either paramagnetic or diamagnetic inductive capacity. There is potential energy of chemical force between any two substances which have what is called affinity for one another,—for instance, between fuel and oxygen, between food and oxygen, between zinc in a galvanic battery and oxygen. There is potential energy of chemical force among the different ingredients of gunpowder or gun cotton. There is potential energy of what may be called chemical force, among the particles of soft phosphorus, which is spent in the allotropic transformation into red phosphorus ; and among the particles of prismatically crystallised sulphur, which is spent when the substance assumes the octohedral crystallisation.

To make chemical combination take place without generating its equivalent of heat, all that is necessary is to resist the chemical force operating in the combination, and take up its effect in some other form of energy than heat. In a series of admirable researches on the agency of electricity in transformations of energy,[1] Joule showed that

[1] "On the Production of Heat by Voltaic Electricity," communicated to the Royal Society December 17th, 1840 (see *Proceedings* of that date), and published *Phil. Mag.* October, 1841.

"On the Heat evolved by Metallic Conductors of Electricity, and in the cells of a battery during Electrolysis."—*Phil. Mag.* October, 1841.

"On the Electrical Origin of the Heat of Combustion."—*Phil. Mag.* March, 1843.

"On the Heat evolved during the Electrolysis of Water," *Proceedings of the Literary and Philosophical Society of Manchester,* 1843, Vol. vii. Part 3, Second Series.

"On the Calorific Effects of Magneto-Electricity, and on the Mechanical Value of Heat," communicated to the British Associa-

the chemical combinations taking place in a galvanic battery may be directed to produce a large, probably in some forms of battery an unlimited, proportion of their heat, not in the locality of combination, but in a metallic wire at any distance from that locality ; or that they may be directed not to generate that part of their heat at all, but instead to raise weights, by means of a rotating engine driven by the current. Thus if we allow zinc to combine with oxygen by the beautiful process which Grove has given in his battery, we find developed in a wire connecting the two poles the heat which would have appeared directly if the zinc had been burned in oxygen gas ; or if we make the current drive a galvanic engine, we

tion (Cork), August 1843, and published *Phil. Mag.* October, 1843.

"On the Intermittent Character of the Voltaic Current in certain cases of Electrolysis, and on the Intensity of various Voltaic arrangements."—*Phil. Mag.* February, 1844.

"On the Mechanical Powers of Electro-Magnetism, Steam, and Horses." By Joule and Scoresby.—*Phil. Mag.* June, 1846.

"On the Heat disengaged in Chemical Combination."—*Phil. Mag.* June 1852.

"On the Economical Production of Mechanical Effect from Chemical Forces."—*Phil. Mag.* Jan. 1853.

All these articles are republished in vol. i. of Joule's *Collected Papers.*

have in weights raised, an equivalent of potential energy for the potential energy between zinc and oxygen spent in the combination.

The economic relations between the electric and the thermo-dynamic method of transformation from chemical affinity to available motive power were indicated, in accordance with the limited capability of heat to be transformed into potential energy, which the modification of Carnot's principle, previously alluded to, shows; and the unlimited performance of a galvanic engine in raising weights to the full equivalent of chemical force used, which Joule has established.

The transformation of motive power into light, which takes place when work is spent in an extremely concentrated generation of heat, was referred to. It was illustrated by the ignition of platinum wire by means of an electric current driven through it by the chemical force between zinc and oxygen in the galvanic battery; and by the ignition and volatilisation of a silver wire by an electric current driven through it by the potential energy laid up in a Leyden battery, when charged by an electrical machine. The luminous heat generated in the last-mentioned case was the complement to a deficiency of heat of friction in the plate-glass and rubber of the machine, which a perfect determination, and comparison with the amount of work spent in turning the machine, would certainly have detected.

The application of mechanical principles to the mechanical actions of living creatures was pointed out. It appears certain, from the most careful physiological researches, that a living animal has not the power of originating mechanical energy ; and that all the work done by a living animal in the course of its life, and all the heat that has been emitted from it, together with the heat that would be obtained by burning the combustible matter which has been lost from its body during its life, and by burning its body after death, make up together an exact equivalent to the heat that would be obtained by burning as much food as it has used during its life, and an amount of fuel that would generate as much heat as its body if burned immediately after birth.

On the other hand, the dynamical energy of luminiferous vibrations was referred to as the mechanical power allotted to plants (not mushrooms or funguses, which can grow in the dark, are nourished by organic food like animals, and like animals absorb oxygen and exhale carbonic acid), to enable them to draw carbon from carbonic acid, and hydrogen from water.

In conclusion, the sources available to man for the production of mechanical effect were examined and traced to the sun's heat and the rotation of the earth round its axis.

Published speculations[1] were referred to, by which it is shown to be possible that the motions of the earth and of the heavenly bodies, and the heat of the sun, may all be due to gravitation ; or *that the potential energy of gravitation may be in reality the ultimate created antecedent of all motion, heat, and light at present existing in the universe.*[1]

[1] *Trans. Roy. Soc.* Edinburgh, April, 1854 Professor W. Thomson, "On the Mechanical Energies of the Solar System." Also *British Association Report*, Liverpool, September, 1854 "On the Mechanical Antecedents of Motion, Heat, and Light." (Republished in Vol. II. of Mathematical and Physical Papers).

ON THE SOURCES OF ENERGY IN NATURE AVAILABLE TO MAN FOR THE PRODUCTION OF MECHANICAL EFFECT.

[*From Presidential Address to the Mathematical and Physical Section of the British Association, York,* 1881.]

DURING the fifty years' life of the British Association, the Advancement of Science for which it has lived and worked so well, has not been more marked in any department than in one which belongs very decidedly to the Mathematical and Physical Section—the science of Energy. The very name energy, though first used in its present sense by Dr. Thomas Young about the beginning of this century, has only come into use practically after the doctrine which defines it had, during the first half of the British Association's life, been raised from a mere formula of mathematical dynamics to the position it now holds of a principle pervading all nature and guiding the investigator in every field of science.

A little article communicated to the Royal Society of Edinburgh a short time before the commencement of the epoch of energy under the title " On the Sources Available to Man for the Production of Mechanical Effect "[1] contained the following :—

" Men can obtain mechanical effect for their
" own purposes by working mechanically them-
" selves, and directing other animals to work for
" them, or by using natural heat, the gravitation
" of descending solid masses, the natural motions
" of water and air, and the heat, or galvanic
" currents, or other mechanical effects produced
" by chemical combination, but in no other way
" at present known. Hence the stores from
" which mechanical effect may be drawn by
" man belong to one or other of the following
" classes :—

[1] Read at the Royal Society of Edinburgh on February 2, 1852 : Published in *Proceedings* of that date ; and republished in Vol. I. of Mathematical and Physical Papers.

" I. The food of animals.

" II. Natural heat.

" III. Solid matter found in elevated posi-
" tions.

" IV. The natural motions of water and air.

' V. Natural combustibles (as wood, coal
" coal-gas, oils, marsh-gas diamond, native sul-
" phur, native metals, meteoric iron).

" VI. Artificial combustibles (as smelted or
" electrically-deposited metals, hydrogen, phos-
" phorus).

" In the present communication, known facts
" in natural history and physical science, with
" reference to the sources from which these stores
" have derived their mechanical energies, are
" adduced to establish the following general con-
" clusions :—

" 1. *Heat radiated from the sun* (sunlight being
" included in this term) *is the principle source of*
" *mechanical effect available to man.*[1] From it is
" derived the whole mechanical effect obtained by

[1] A general conclusion equivalent to this was published by Sir John Herschel in 1833. See his *Astronomy*, edit. 1849, § (399).

F F 2

" means of animals working, water-wheels worked
" by rivers, steam-engines, galvanic engines, wind-
" mills, and the sails of ships.

" 2. The motions of the earth, moon, and sun,
" and their mutual attractions, constitute an im-
" portant source of available mechanical effect.
" From them all, but chiefly no doubt from the
" earth's motion of rotation, is derived the me-
" chanical effect of water-wheels driven by the
" tides.

" 3. The other known sources of mechanical
" effect available to man are either terrestrial—
" that is, belonging to the earth, and available
" without the influence of any external body—or
" meteoric—that is, belonging to bodies deposited
" on the earth from external space. Terrestrial
" sources, including mountain quarries and mines,
" the heat of hot springs, and the combustion
" of native sulphur, perhaps also the combustion
" of inorganic native combustibles, are actually
" used ; but the mechanical effect obtained from
" them is very inconsiderable, compared with
" that which is obtained from sources belonging

" to the two classes mentioned above. Meteoric
" sources, including only the heat of newly-fallen
" meteoric bodies, and the combustion of meteoric
" iron, need not be reckoned among those avail-
" able to man for practical purposes."

Thus we may summarise the natural sources of
energy as Tides, Food, Fuel, Wind and Rain.

Among the practical sources of energy thus
exhaustively enumerated, there is only one not derived from sun-heat—that is the tides. Consider
it first. I have called it *practical*, because tidemills exist. But the places where they can work
usefully are very rare, and the whole amount of
work actually done by them is a drop to the
ocean of work done by other motors. A tide of
two meters' rise and fall, if we imagine it utilised
to the utmost by means of ideal water-wheels
doing with perfect economy the whole work of
filling and emptying a dock-basin in infinitely
short times at the moments of high and low water,
would give just one metre-ton per square metre
of area. This work done four times in the
twenty-four hours amounts to 1-1620th of the

work of a horse-power. Parenthetically, in explanation, I may say that the French metrical equivalent (to which in all scientific and practical measurements we are irresistibly drawn, notwithstanding a dense barrier of insular prejudice most detrimental to the islanders),—the French metrical equivalent of James Watt's "horse-power" of 550 foot-pounds per second, or 33,000 foot-pounds per minute, or nearly two million foot-pounds per hour, is 75 metre-kilogrammes per second, or $4\frac{1}{2}$ metre-tons per minute, or 270 metre-tons per hour. The French ton of 1,000 kilogrammes used in this reckoning is 0·984 of the British ton.

Returning to the question of utilising tidal energy, we find a dock area of 162,000 square metres (which is little more than 400 metres square) required for 100 horse-power. This, considering the vast costliness of dock construction, is obviously prohibitory of every scheme for economising tidal energy by means of artificial dock-basins, however near to the ideal perfection might be the realised tide-mill, and however convenient and non-wasteful the accumulator—whether Faure's electric accumulator, or other accumulators of energy hitherto invented or to be invented—which might be used to store up the energy yielded by the tide-mill during its short harvests about the times of high and low water, and to give it out when wanted at other times of six hours. There may, however, be a dozen places possible in the world where it could be advantageous to build a sea-wall across the mouth of a natural basin or estuary, and to utilise the tidal energy of filling it and emptying it by means of sluices and water-wheels. But if *so* much could be done, it would in many cases take only a little more to keep the water out altogether, and make fertile land of the whole basin. Thus we are led up to the interesting economical question, whether is forty acres (the British *agricultural* measure for the area of 162,000 square metres) or 100 horse-power more valuable. The annual cost of 100 horse-power night and day, for 365 days of the year, obtained through steam from coals, may be about ten times the rental of forty acres at 2*l*. or 3*l*. per acre. But the value of land is essentially much more than its rental, and the rental of land is apt to be much more than 2*l*. or 3*l*. per acre in places where 100 horse-power could be taken with advantage from coal through steam. Thus the question remains unsolved, with the possibility that in one place the answer may be *one hundred horse-power*, and in another *forty acres*. But, indeed, the question is hardly worth answering, considering the rarity of the cases, if they exist at all, where embankments for the utilisation of tidal energy are practicable.

Turning now to sources of energy derived from sun-heat, let us take the wind first. When we look at the register of British shipping and see 40,000 vessels, of which about 10,000 are steamers and 30,000 sailing ships, and when we think how vast an absolute amount of horse-power is developed by the engines of those steamers, and how considerable a proportion it forms of the whole horse-power taken from coal annually in the whole world at the present time, and when we consider the sailing ships of other nations, which must be reckoned in the account, and throw in the little item of windmills, we find that, even in the present days of steam ascendency, old-fashioned Wind still supplies a large part of all the energy used by man. But however much we may regret the time when Hood's young lady, visiting the fens of Lincolnshire at Christmas, and writing to her dearest friend in London (both sixty years old now if they are alive), describes the delight of sitting in a bower and looking over the wintry plain, not desolate, because " windmills lend revolving animation to the scene," we cannot shut our eyes to the fact of a lamentable decadence of wind-power. Is this decadence permanent, or may we hope that it is only temporary ? The subterranean coal-stores of the world are becoming exhausted surely, and not slowly, and the price of coal is upward bound —upward bound on the whole, though no doubt it will have its ups and downs in the future as it has had in the past, and as must be the case in respect to every marketable commodity. When the coal is all burned ; or, long before it is all burned, when there is so little of it left and the coal-mines from which that little is to be excavated are so distant and deep and hot that its price to the consumer is greatly higher than at present, it is most probable that windmills or wind-motors in some form will again be in the ascendant, and that wind will do man's mechanical work on land at least in proportion comparable to its present doing of work at sea.

Even now it is not utterly chimerical to think of wind superseding coal in some places for a very important part of its present duty—that of giving light. Indeed, now that we have dynamos and Faure's accumulator, the little want to let the thing be done is cheap windmills. A Faure cell containing twenty kilogrammes of lead and minium charged and employed to excite incandescent vacuum-lamps has a light-giving capacity of sixty-candle hours (I have found considerably more in experiments made by myself; but I take sixty as a safe estimate). The charging may be done uninjuriously, and with good dynamical economy, in any time from six hours to twelve or more. The drawing off of the charge for use may be done safely, but somewhat wastefully, in two hours, and very economically in any time of from five hours to a week or more. Calms do not last often longer than three or four days at a time. Suppose, then, that a five days' storage-capacity suffices (there may be a little steam-engine ready to set to work at any time after a four-days' calm, or the user of the light may have a few candles or oil-lamps in reserve, and be satisfied with them when the wind fails for more than five days). One of the twenty kilogramme cells charged when the windmill works for five or six hours at any time, and left with its sixty-candle hours' capacity to be used six hours a day for five days, gives a two-candle light. Thus thirty-two such accumulator cells so used would give as much light as four burners of London sixteen-candle gas. The probable cost of dynamo and accumulator does not seem fatal to the plan, if the windmill could be had for something comparable with the prime cost of a steam-engine capable of working at the same horse-power as the windmill when in good action. But windmills as hitherto made are very costly machines ; and it does not seem probable that, without inventions not yet made, wind can be economically used to give light in any considerable class of cases, or to put energy into store for work of other kinds.

Consider, lastly, rain-power. When it is to be had in places where power is wanted for mills and factories of any kind, water-power is thoroughly appreciated. From time immemorial, water-motors have been made in large variety for utilising rain-power in the various conditions in which it is presented, whether in rapidly-flowing rivers, in natural waterfalls, or stored at heights in natural lakes or artificial reservoirs. Improvements and fresh inventions of machines of this class still go on ; and some of the finest principles of mathematical hydrodynamics have, in the lifetime of the British Association, and, to a considerable degree with its assistance, been put in requisition for perfecting the theory of hydraulic mechanism and extending its practical applications.

A first question occurs : Are we necessarily

limited to such natural sources of water-power as are supplied by rain falling on hill-country, or may we look to the collection of rain-water in tanks placed artificially at sufficient heights over flat country to supply motive power economically by driving water-wheels? To answer it: Suppose a height of 100 metres, which is very large for any practicable building, or for columns erected to support tanks; and suppose the annual rainfall to be three-quarters of a metre (thirty inches). The annual yield of energy would be seventy-five metre-tons per square metre of the tank. Now one horse-power for 365 times 24 hours is 2,365,000 metre-tons; and therefore (dividing this by 75) we find 31,530 square metres as the area of our supposed tank required for a continuous supply of one horse-power. The prime cost of any such structure, not to speak of the value of the land which it would cover, is utterly prohibitory of any such plan for utilising the motive power of rain. We may or may not look forward hopefully to the time when windmills will again "lend revolving animation" to a dull flat country; but we certainly need not be afraid that the scene will be marred by forests of iron columns taking the place of natural trees, and gigantic tanks overshadowing the fields and blackening the horizon.

To use rain-power economically on any considerable scale we must look to the natural drainage of hill country, and take the water where we find it either actually falling or stored up and ready to fall when a short artificial channel or pipe can be provided for it at moderate cost. The expense of aqueducts, or of underground water-pipes, to carry water to any great distance—any distance of more than a few miles or a few hundred yards—is much too great for economy when the yield to be provided for is *power ;* and such works can only be undertaken when the *water itself* is what is wanted. Incidentally, in connection with the water supply of towns, some part of the energy due to the head at which it is supplied may be used for power. There are, however, but few cases (I know of none except Greenock) in which the energy to spare over and above that devoted to bringing the water to where it is wanted, and causing it to flow fast enough for convenience at every opened tap in every house or factory, is enough to make it worth while to make arrangements for letting the water-power be used without wasting the water-substance. The cases in which water-power is taken from a town supply are generally very small, such as working the bellows of an organ, or " hair-brushing by machinery," and involve simply throwing away the used water. The cost of energy thus obtained must be something enormous in proportion to the actual quantity of the energy, and it is only the smallness of the quantity that allows the convenience of having it when wanted at any moment, to be so dearly bought.

For anything of great work by rain-power, the water-wheels must be in the place where the water supply with natural fall is found. Such places are generally far from great towns, and the time is not yet come when great towns grow by natural selection beside waterfalls, for power; as they grow beside navigable rivers, for shipping. Thus hitherto the use of water-power has been confined chiefly to isolated factories which can be conveniently placed and economically worked in the neighbourhood of natural waterfalls. But the splendid suggestion made about three years ago by Mr. Siemens (Sir William Siemens) in his presidential address to the Institution of Mechanical Engineers, that the power of Niagara might be utilised, by transmitting it electrically to great distances, has given quite a fresh departure for design in respect to economy of rain-power. From the time of Joule's experimental electro-magnetic engines developing 90 per cent. of the energy of a Voltaic battery in the form of weights raised, and by the theory of the electro-magnetic transmission of energy completed thirty years ago on the foundation afforded by the train of experimental and theoretical investigations which established his dynamical equivalent of heat in mechanical, electric, electro-chemical, chemical, electro-magnetic, and thermoclastic phenomena, it had been known that potential energy from any available source can be transmitted electro-magnetically by means of an electric current through a wire, and directed to raise weights at a distance, with unlimitedly perfect economy.

The first large-scale practical application of electro-magnetic machines was proposed by Holmes
in 1854, to produce the electric light for light-
houses, and persevered in by him till he proved
the availability of his machine to the satisfaction
of the Trinity House and the delight of Faraday
in trials at Blackwall in April, 1857, and it was
applied to light the South Foreland lîghthouse on
December 8, 1858. This gave the impulse to invention ; by which the electro-magnetic machine
has been brought from the physical laboratory into
the province of engineering, and has sent back to
the realm of pure science a beautiful discovery—
that of the fundamental principle of the dynamo,
made triply and independently, and as nearly as
may be simultaneously, in 1867 by Dr. Werner
Siemens, Mr. S. A. Varley, and Sir Charles Wheat-
stone ; a discovery which constitutes an electro-
magnetic analogue to the fundamental electrostatic
principle of Nicholson's revolving doubler, resuscitated by Mr. C. F. Varley in his instrument " for
generating electricity" ; patented in 1860 ; and
by Holtz in his celebrated electric machine ; and

by myself in my "replenisher" for multiplying and maintaining charges in Leyden jars for heterostatic electrometers, and in the electrifier for the siphon of my recorder for submarine cables.

The dynamos of Gramme and Siemens, invented and made in the course of these fourteen years since the discovery of the fundamental principle, give now a ready means of realising economically on a large scale, for many important practical applications, the old thermo-dynamics of Joule in electro-magnetism ; and, what particularly concerns us now in connection with my present subject, they make it possible to transmit electro-magnetically the work of waterfalls through long insulated conducting wires, and use it at distances of fifties or hundreds of miles from the source, with excellent economy—better economy, indeed, in respect to proportion of energy used to energy dissipated than almost anything known in ordinary mechanics and hydraulics for distances of hundreds of yards instead of hundreds of miles.

ON THE DISSIPATION OF ENERGY

[*Being article in* " *The Fortnightly Review*," *March*, 1892.]

THE old chimera of "the perpetual motion" still lives, not so much in popular belief as in the scientific imagination. If we are now to feel sure that it has no more real existence than the fabled monster of Lycia and Etna, it is primarily because naturalists have failed, after diligent and persevering search, with all the help they could get from the science and art department of mankind ever since its commencement many thousand years ago, to find any creature fulfilling the imagined characteristics ; not because philosophy can prove any absurdity in the idea that such a species should exist. In its original form of a machine which could do work without food, or fuel, or supply of energy from wind or water, or other external source, the perpetual motion was dead to science long before Newton's time: and on the negation of it Stevinus founded a beautiful proof of the parallelogram of forces, which is celebrated in the history of dynamics, and is still justly admired. But the doctrine of the "Conservation of Energy," which has grown up since the end of last century, has given a fresh lease of life to the idea of the perpetual motion revived in a more subtle form.

From Rumford, Davy, and Joule we have learned that the reason why every machine, even though not called upon to give out work done by it, must come to rest, is not, as was generally supposed by contemporary and preceding philosophers, because the friction that stops the machine implies annihilation of energy, but because it converts into heat the energy given initially in the motion of the machine. Suppose now we could guard perfectly against loss of heat by radiation, or by cooling currents of air, or by conduction along the supports of the machine, might we not annex to it a motor, acting on the same principle as the steam-engine, which would reconvert into motion of the machine the heat which is developed by friction? Have we not here a good scientific foundation for believing that a fly-wheel set in motion, or clock-work driven by the unwinding of a spring or the running down of a weight, and connected with a heat engine worked by the heat generated by its friction, only wants an impermeable encloser preventing all loss of heat to allow it to go on for ever? Of course, this impermeable encloser is not realisable, but it is both a scientific and a practical consideration to think what might be done if we had an impermeable substance of which an enclosing case for the instrument could be constructed. We know by the principle of the "Conservation of Energy" that all the energy we gave to the machine is always all there; some of it in heat and the rest in energy of the weight or spring not quite run down, or in the visible motion of the fly-wheel, or wheels, or vibrating pendulum, or other moving parts of the mechanism.

Why not convert and re-convert continually into motion of the fly-wheel, or energy of the spring, or weight wound up, all the heat generated by the friction in the machine? To this question Carnot,[1] in 1824, in his *Réflexions sur la Puissance Motrice du Feu*, showed how to find a negative answer, to be founded, not on any then known

[1] Sadi Carnot, born in 1796, son of the Republican War-Minister, and uncle of the present President of the French Republic. He inherited from his father a chivalrous motivity of disposition, which was prettily illustrated by a little piece of history of the year 1800 told by his brother Hippolyte, in the biographical sketch referred to below.

The Directory had been superseded by the Consulate. Carnot having returned to his country after two years of exile, was called to be War Minister. . . . When the Minister went to the Malmaison for his official work with the first Consul he often brought with him his son, about four years old. The boy on these occasions lived with Madame Bonaparte, who had a great affection for him. One day she was rowing about in a boat with some of her ladies. Bonaparte came and amused himself by throwing stones into the water round the boat, so as to splash the fresh dresses of the rowers. The ladies did not dare to show their displeasure openly. The little boy, after having watched for some time what was going on, came suddenly and squared up to the conqueror of Marengo, threatening him with his fist, and cried out, "*Animal de Premier Consul, veux-tu ne pas taquiner ces dames!*" Bonaparte at this unexpected attack stopped, looked with astonishment at the child, and then fell into a hearty fit of laughter which spread to all the spectators of the scene.

law or principle in Natural Philosophy, but rather on general observation of natural phenomena, on experience in practical mechanics, and on experimental investigation of properties of matter ;—an answer founded on knowledge acquired in what may be called the "natural history stage" of progress towards truth.

That little essay was indeed an epoch-making gift to science. From it we have learned that heat is only available for a steam-engine, or an air-engine, or a gas-engine, in proportion to the excess of the temperature of the matter in which it is given above the temperature of the coldest matter obtainable for use in connection with the engine to carry heat away from it continually during the time it is working.

Every heat motor (as for brevity we may call any heat engine doing mechanical work in virtue of heat supplied to it) requires difference of temperature in different parts; or in the same part at different times, as in the old Newcomen condensing-engine before Watt's improvement of the separate condenser was introduced. Heat is

essentially taken in by the engine at the higher
temperature and given out at the lower temperature. All this was taught by Carnot, in 1824,
but with it, in his original essay, was involved the
then almost universally prevailing idea that heat
was a material substance, and that therefore the
quantity of heat given out by the engine at the
lower temperature must be exactly equal to the
quantity of heat taken in at the higher temperature.
Carnot died in 1832 (two years after the Revolution of 1830), at the age of thirty-six. If he had
lived a few years longer, or if his short life, begun
in the Reign of Terror, had been less troubled [1]

---

[1] "These researches" [in thermodynamics] "were roughly
interrupted by a great event, the Revolution of July, 1830. . . .
Sadi frequented the popular meetings of this epoch, without, however,
going beyond the character of a simple observer. . . . On the day of
the funeral of General La Marque, Sadi was taking a walk out
of curiosity in the neighbourhood of the insurrection. A mounted
soldier, who seemed drunk, passed at a gallop through the street
brandishing his sabre and striking at passers-by. Sadi dashed
forward, skilfully avoided the weapon of the soldier, seized him by
the leg, dragged him off his horse, laid him gently in the gutter,
and continued his walk ; stealing himself away from the acclama-
tions of the crowd, who were astonished at this bold *coup de main*."
—From *Notice Biographique*, p. 78, by his brother Hippolyte
Carnot, referred to below.

by the political miscarriages of his country and repetitions of revolutionary violence, we should have learned much more from him. Manuscript journals and memorandums, found among his papers and published [1] after his death (but not published before Joule had finally convinced the world of the immateriality of heat and had measured its dynamical equivalent), proved that Carnot had lived long enough to see irrefragable reasons for abandoning the doctrine of the materiality of heat and for confidently believing that heat is in reality motion among the particles or molecules or atoms of matter ; and that he had taught himself decisively and thoroughly the doctrine of the "Conservation of Energy," which ten years later was given to the world by Joule

[1] "*Réflexions sur la Puissance Motrice du Feu et sur les Machines propre a Développer cette Puissance, par S. Carnot, ancien Élève de l'École Polytechnique*, Paris, 1878. Of this publication, with its appendices of biographical sketch by his younger brother Hippolyte Carnot, and extracts from unpublished writings of Sadi, an English version has been published in America (and in England, Macmillan & Co., 1890) under the editorship of Dr. Thurston, Cornell University, who adds to it a short article by himself, on "The Work of Sadi Carnot," full of interesting matter.

with his first determination of the Mechanical Equivalent of Heat.

To the reprint (sixty-five pages) of Carnot's original essay of 1824 are appended thirty-three pages of *Extrait de Notes Inédites de Sadi Carnot, sur les Mathématiques, la Physique, et autres sujets*, and twenty-one pages of biographical sketch of the author, by his younger brother, Hippolyte Carnot, whose name, as a very benevolent writer and worker in political and social affairs, was well known in 1845[1] among Paris booksellers, none of whom, so far as my inquiries went, had ever heard of Sadi or his *Réflexions sur la Puissance Motrice du Feu.*

Here are some of Carnot's words literally translated (from pp. 95, 96):—

"Heat is nothing else than motive power, or "rather motion which has changed its form. It

---

[1] I went to every book-shop I could think of, asking for the *Puissance Motrice du Feu*, by Carnot. "Caino? Je ne connais pas cet auteur." With much difficulty I managed to explain that it was "r" not "i" I meant. "Ah! Ca-rrr-not! Oui, voici son ouvrage," producing a volume on some social question by Hippolyte Carnot; but the *Puissance Motrice du Feu* was quite unknown.

" is a motion among the particles of bodies.
" Wherever there is destruction of motive power
" there is at the same time production of heat in
" quantities precisely proportional to the quantity
" of motive power destroyed. Conversely wherever
" there is destruction of heat there is production
" of motive power.

" We may then assert the general proposition
" that motive power is of invariable amount in
" nature ; that it can never, properly speaking, be
" said to be either produced or destroyed. In
" truth, it experiences changes of form, that is to
" say, it produces sometimes one kind of movement,
" and sometimes another, but it is never annulled."

These words contain a perfectly clear and
general statement of the " Conservation of
Energy ; " but Carnot did not live long enough
to see how his original doctrine of the motive
power of fire was to be reconciled to this principle.
He says (p. 92) :—

" It would be difficult to say why, in the develop-
" ment of motive power by consuming the heat of
" a hot body, a cold body is necessary ; or why we

" cannot produce motion simply by consuming the
" heat of a hot body."

" When we produce motive power by the passage
" of heat from the body A to the body B, is the
" quantity of this heat which is delivered to B (if it
' is not of the same amount as that taken from A,
" if a part is really consumed to produce motive
" power) the same, whatever be the substance
" employed [in the ideal engine] to realise the
" motive power ? "

" Could there be possibly a means [or substance]
" for causing more heat to be consumed in pro-
" ducing motive power, and, therefore, less to be
" delivered to the body B ? Would it be possible
" even to consume the whole heat taken from A,
" without the necessity of delivering any heat to,
" B ? *If this were possible we could create motive
" power without fuel, and simply by destruction of
" some of the heat of bodies.*"

In these last words (which I have given in
italics) we have from the founder of our theory of
the steam-engine and other heat motors, and the
profoundest thinker in thermodynamic philosophy

of the first thirty years of the nineteenth century, a thoroughly clear statement of the old perpetual motion in its most subtle nineteenth-century form. But this statement is put as a question with clear indication of a bias towards a negative answer : and it is impossible to doubt that Carnot would have unhesitatingly given the negative answer if a little more time had been allowed him for thinking out the thermodynamic problem. Happily, however, Carnot's original essay led others to give it. My brother, Professor James Thomson, assumed a negative answer, and founded on it his theoretical demonstration that the freezing point of water is lowered by pressure.[1]

Two years later [2] I gave the negative answer as

---

[1] *Transactions of the Royal Society of Edinburgh*, January 2nd 1849, reprinted in *Cambridge and Dublin Mathematical Journal*, November, 1850, and quoted *in extenso* in vol. i., *Mathematical and Physical Papers*, Sir W. Thomson (pp. 156—164).

[2] *Transactions of the Royal Society of Edinburgh*, March, 1851, and *Philosophical Magazine*, iv. 1852, " On the Dynamical Theory of Heat, with Numerical Results deduced from Mr. Joule's Equivalent of a Thermal Unit, and M. Regnault's Observations on Steam," reprinted in vol. i., Sir W. Thomson's *Mathematical and Physical Papers*.

an axiom in the following terms :—" It is impossible, by means of inanimate material agency, to derive mechanical effect from any portion of matter by cooling it below the temperature of the coldest of the surrounding objects. If this axiom be denied for all temperatures, it would have to be admitted that a self-acting machine might be set to work and produce mechanical effect by cooling the sea or earth, with no limit but the total loss of heat from the earth and sea, or, in reality, from the whole material world."

My statement of this axiom was limited to inanimate matter because not enough was known either from the natural history of plants and animals or from experimental investigations in physiology to assert with confidence that in animal or vegetable life there may not be a conversion of heat into mechanical effect not subject to the conditions of Carnot's theory. It seemed to me then, and it still seems to me, most probable that the animal body does not act as a thermodynamic engine in converting heat produced by the combination of the food with the oxygen of the inhaled air, but that it acts in a manner more nearly analogous to that of an electric motor working in virtue of energy supplied to it by a voltaic battery. According to either view, however, the mechanical effect achieved by an animal in walking up-hill, or in flying or swimming, or in dragging loads along the ground, or in acting as motor for a horse-mill, or tread-mill, or a crank, or a lever as for pumping, or for any kind of mechanism, is a part equivalent for the oxidation of the food ; the rest of the equivalent being animal heat. Joule estimated that from $\frac{1}{4}$ to $\frac{1}{8}$ of the dynamical equivalent of the complete oxidation of all the food consumed by a horse may be produced from day to day in mechanical effect as of weights raised, the remainder, or from $\frac{3}{4}$ to $\frac{5}{8}$, being evolved and given out as heat ; and similar proportions seem to hold for the mechanical work and the development of heat by a healthy vigorous working man. It is, however, conceivable that animal life might have the attribute of using the heat of surrounding matter, at its natural temperature, as a source of energy for mechanical effect, and thus constituting a case of affirmative answer for Carnot's last thermodynamic question. The influence of animal or vegetable life on matter [1] is infinitely beyond the range of any scientific inquiry hitherto entered on. Its power of directing the motions of moving particles, in the demonstrated daily miracle of our human free-will, and in the growth of generation after generation of plants from a single seed, are infinitely different from any possible result of the fortuitous concourse of atoms ; and *the fortuitous concourse of atoms is the sole foundation in Philosophy on which can be founded the doctrine that it is impossible to derive mechanical effect from heat otherwise than by taking heat from a body at a higher temperature, converting at most a definite proportion of it into mechanical effect, and giving out the whole residue to matter at a lower temperature.*

The considerations of ideal reversibility, by

---

[1] About twenty-five years ago, I asked Liebig if he believed that a leaf or a flower could be formed or could grow by chemical forces. He answered, "I would more readily believe that a book on chemistry or on botany could grow out of dead matter by chemical processes."

which Carnot was led to his theory, and the true reversibility of every motion in pure dynamics have no place in the world of life. Even to think of it (and on the merely dynamical hypothesis of life we can think of it as understandingly as of the origination of life and evolution of living beings without creative power), we must imagine men, with conscious knowledge of the future but with no memory of the past, growing backward and becoming again unborn; and plants growing downwards into the seeds from which they sprang. But the real phenomena of life infinitely transcend human science: and speculation regarding consequences of their imagined reversal is utterly unprofitable. Far otherwise, however, it is in respect of the reversal of the motions of matter uninfluenced by life, a very elementary consideration of which leads to the full explanation of the theory of the dissipation of energy.

Carnot's theory of the perfect heat engine is essentially founded on the consideration of a reversible cycle of processes. The perfect engine is essentially an engine which can be worked backwards with every action in its cycle exactly reversed. When working forwards it performs mechanical work in virtue of heat taken from a hot body, A, of which a certain portion is essentially given to a body, B, at a lower temperature. To reverse its action mechanical work must be done upon it, and the equivalent output is a certain quantity of heat taken from the cold body, B, and a greater quantity given to the hot body, A. The excess of the quantity of heat taken from A above that given to B when the engine works forwards, and the excess of the heat given to A above that taken from B when the engine is worked backwards, is equal to the quantity of heat which has the same dynamical energy as the work done *by the engine*, in the case of working forwards and the work done *upon the engine by an external, agent*, when the engine is worked backwards.

It is impossible to fulfil the condition of perfect reversibility by any engine composed of any real material to be found in nature. The friction of the parts, and the impossibility of getting heat into the engine from A, and causing heat to leave the engine and pass into B, except by falls of temperature from the temperature of A to the highest effective temperature of the engine, and from the lowest effective temperature in the engine to the temperature of B, violate the condition of perfect reversal and involve essentially irreversible actions in the cycle of the engine, whether working forwards or worked backwards. In the condensing steam-engine, A is the burning coal of the furnace. The highest effective temperature in the engine is the temperature of the steam entering the cylinder from the boiler. The lowest effective temperature is the temperature of the "exhaust steam," that is to say, of the steam coming out of the cylinder in a single cylinder engine, or out of the lowest-pressure cylinder in a triple or quadruple expansion engine. In a condensing engine, B is the condensing water: in the non-condensing engine B is the air into which the waste steam is blown. The superiority of the double, triple, and quadruple expansion engines, over a single cylinder engine, is due to their diminishing the ineffective

H H 2

droppings down of temperature, between the highest temperature to which the water of the boiler can be raised for safe and effective use, and the temperature of the exhaust steam. The superior efficiency of a condensing engine consists in its allowing the temperature of the exhaust steam to be about 40° or 50° C., instead of its being a degree or two above 100°, as it essentially is in the non-condensing expansive engine. James Watt was, by his separate condenser, his use of expansion in single cylinder engines, and his origination of the now generally employed plan of double, or triple, or quadruple expansion engine, with his perfect tact and judgment as to practical economy, and his profound scientific knowledge of mechanics and of the properties of steam, arranging his engine to as nearly as possible fulfil Carnot's condition of reversibility, by minimising every irreversible action in its cycle of work. But it seems certain that he had no idea of Carnot's grand generalisation, according to which one perfectly reversible engine would give exactly as much work as any other, of whatever different substance or character, using heat supplied at the same temperature, and having the same lower temperature available for the carrying away of waste heat.

Exhaustive consideration of all that is known of the natural history of the properties of matter, and of all conceivable methods for obtaining mechanical work from natural sources of energy, whether by heat engines, or electric engines, or water-wheels, or windmills, or tidemills, or any other conceivable kind of engine, proves to us that the most perfectly designed engine can only be an approach to the perfect engine ; and that the irreversibility of actions connected with its working is only part of a physical law of irreversibility according to which there is a universal tendency in nature to the dissipation of mechanical energy ; and any partial *restoration* of mechanical energy is impossible in inanimate material processes, and is probable never effected by means of organised matter, either endowed with vegetable life, or subject to the will of an animal.

Some mathematical details regarding cases of

this law will be found in a short paper[1] in the *Proceedings of the Royal Society of Edinburgh* for April 19, 1852. The dynamical explanation of it, founded essentially on consideration of the vastness of the numbers of freely moving atoms or particles in even the smallest portion of palpable matter, and the infinity of such motions in the material universe, is given in a paper entitled " The Kinetic Theory of the Dissipation of Energy," which was communicated to the Royal Society of Edinburgh, twenty-two years later,[2] and which is republished in the *Philosophical Magazine* for the present month (March, 1892).

We have been considering a fly-wheel or clock-work driven by a weight and the heat generated by friction against the motion of wheels and pendulum, and by impacts of teeth against the pallets of an escapement. Our knowledge of properties of matter and of modes of propagation of heat by

[1] "On a Universal Tendency in Nature to the Dissipation of Mechanical Energy," republished in vol. i. of *Mathematical and Physical Papers*, pp. 511—514.

[2] *Proceedings of the Royal Society of Edinburgh* February 16, 1874.

radiation or conduction, and of tne efficiency of heat as a motor, discovered by several thousand years of observation and several hundred years of experiment and dynamical theory, suffices to show that when the weight is run down, and the potential energy (or capacity to do work), which it had in the beginning, has been all spent in heat, this heat is not available for raising the weight and giving the clockwork a renewed lease of motivity. The solar system, according to the best of modern scientific belief, is dynamically analogous to the clockwork, in all the essentials of our consideration. Not going back in thought to a beginning of which science knows nothing, let us compare the solar system as it was three thousand years ago with the solar system as it is now. Let our analogue be a clockwork which three hours ago was known to be going with its weight partially run down, and which is still going with its weight not yet wholly run down.

During these three thousand years the sun has been giving out radiant heat (light being included in the designation " radiant heat ") in all directions, propagated at the rate of about nine and a half million million kilometres [1] per year, and therefore twenty-eight and a half thousand million million kilometres in three thousand years. We do not know for certain whether the light which left the sun three thousand years ago is still travelling outwards with almost undiminished energy, or whether nearly all is already dissipated in heat, warming the luminiferous ether, or ponderable bodies which have obstructed its course. But we may, I think, feel almost sure that it is partly still travelling outwards as radiant heat, and partly spent (or dissipated) in warming ponderable matter (or ponderable matter and the luminiferous ether).

The running down of the weight in the clockwork has its perfect analogue, as Helmholtz was, I believe, in reality the very first to point out, in the shrinkage of the sun from century to century under the influence of the mutual gravitational attractions between its parts. The heat-producing efficiency of the fire which there would be if the sun

---

[1] The "kilometre" is sixty-two hundredths of the British statute mile ; rather a long half mile in fact.

were a globe of gunpowder or guncotton burning from its outward surface inwards—that is to say, the work done by the potential energy of the chemical affinity between uncombined oxygen, and carbon and hydrocarbons, attractive forces as truly forces, and subject to dynamic law, as is the force of gravity itself, is absolutely infinitesimal in comparison with the work done by the gravitational attraction on the shrinking mass adduced by Helmholtz as the real source of the sun's heat.

The whole store of energy now in the sun, whether of actual heat, corresponding to the sun's high temperature, or of potential energy (as of the not run-down weight of the clockwork)—potential energy of gravitation depending on the extent of future shrinkage which the sun is destined to experience, is essentially finite ; and there is much less of it now than there was three hundred thousand years ago. Similar considerations of action on a vastly smaller scale are of course applicable to terrestrial plutonic energy, and thoroughly dispose of the terrestrial " perpetual motion " by which

Lyell[1] and other followers of Hutton, on as sound principles as those of the humblest mechanical perpetual-motionist, tried to find that the earth can go on for ever as it is, illuminated by the sun from infinity of time past to infinity of time future, always a habitation for race after race of plants and animals, built on the ruins of the habitations of preceding races of plants and animals. The doctrine of the "Dissipation of Energy" forces upon us the conclusion that within a finite period of time past the earth must have been, and within a finite period of time to come must again be, unfit for the habitation of man as at present constituted, unless operations have been and are to be, performed which are impossible under the laws governing the known operations going on at present in the material world.

[1] *Principles of Geology*, vol. ii., edition 1868, p. 213 and pp. 40—243 (recapitulation of Chapters xxxi. and xxxiii., 1, 10, 15).

THE BANGOR LABORATORIES

[*Address delivered on the occasion of the opening of the Physical and Chemical Laboratories in University College, Bangor, North Wales, February 2nd,* 1885.]

I FEEL that the present occasion, upon which you have done me the honour to ask me to preside, is one of very great importance indeed, and I wish some person more competent to preside on such an occasion and give a suitable inaugural address were in my place. I am afraid I must confine myself to something not at all worthy of the greatness of an occasion which is almost the opening of a new university. Not quite so, because the real opening of this college took place several months ago; but still it is an occasion which I feel to be much more than merely the opening of a department—a working department—in the college; an occasion of so great moment that I regret that I shall not be able to give anything that could be properly considered a worthy inaugural address. I shall be obliged to ask your indulgence if I confine myself specially to departments with which I am personally familiar—scientific laboratories. The laboratory of a scientific man is his place of work. The laboratory of the geologist and of the naturalist is the face of this beautiful world. The geologist's laboratory is the mountain, the ravine and the seashore. The naturalist and the botanist go to foreign lands, to study the wonders of nature, and describe and classify the results of their observations. But they must do more than merely describe, represent, and depict what they have seen. They must bring home the products of their expeditions to their studies, and have recourse to the appliances of the laboratory properly so-called for their thorough and detailed examination. The naturalist in his laboratory, with his microscope and appliances for the keenest examination, learns to know more than can be learned by merely looking at external beauties. The geologist brings his specimens to the chemist—is himself a chemist perhaps—brings his crystals to the physical laboratory to be examined as to their physical properties, their hardness, the angles between their faces, their optical qualities. Some people might think this an ignoble way to deal with crystals. But it is not so to the trained eye and deeper thought of the scientific man. The scientific man sees and feels beauty as much as any mere observer—as much as any artist or painter. But he also sees something underlying that beauty ; he wishes to learn something of the actions and forces producing those beautiful results. The necessity for study below the surface seems to have been earliest recognised in anatomy, and earliest carried out in human anatomy. I am not going to speak of the work of scientific research generally, but with reference to the special occasion which brings us here this day—the opening of the chemical and physical laboratories of the University College of North Wales. I am going to speak of laboratories for students, laboratories in which the students work with their own hands. There have been laboratories of investigation from the earliest times. No doubt Aristotle had his; and Archimedes had a laboratory wherever he went—in his bath, even, he observed, and studied, and thought out the laws of hydrostatics. But those were not students' laboratories, and our special subject to-day is a students' laboratory, where they can meet together for the practical study of the various departments of science, where they will be brought together to use their eyes and hands— their eyes otherwise than in merely reading books and looking at pictures or drawings; their eyes to observe accurately, and their hands to experiment, in order to learn more than can be learned by mere observation. To teach students to so work and so learn is the object of a scientific student's laboratory.

The first scientific laboratory that ever existed was that of Frederick II., King of Sicily, and was established between 1200 and 1250. Acting under the advice of his chief physician, Martianus, he made a law that nobody should practise physic

or surgery without having studied anatomy practically. He established a school of practical
anatomy, to which students flocked from all parts
of Europe for many years. Subsequently there
was an anatomical school instituted at Bologna ;
and in those two schools we hear the first of
students working in laboratories. The anatomical
students' working-room has for several hundred
years been generally recognised as an absolute
necessity of medical education. But I believe
there was no other branch of physical science
where students worked in the laboratory until
twenty years of the present century had
passed away. The University of Glasgow is, I
think, justly entitled to take some pride in the
great modern expansion and extension of the
system of giving students practical work in
laboratories, as an addition to the education
which previously had been confined almost entirely to book-work, or, at best, to attending
lectures illustrated by experiments and diagrams.
The first chemical laboratory for students, so far
as I know, was that founded by a colleague of my

own name, though no relation—Thomas Thomson [1]
the great chemist and mineralogist. Prior to 1831
a students' chemical laboratory, under Thomas
Thomson, at Glasgow University, flourished and

[1] [Note added February 12, 1885 :—First Professor of Chemistry
in Glasgow University ; appointed 1818 ; held the chair till his
death, 1852.

The minutes of the Faculty of Glasgow College show that as
early as the first month of 1828, Prof. Thomas Thomson began
applying for more commodious premises in which to carry on his
work in the department of chemistry. For two years he kept his
wants persistently before the Faculty (of which he, being only a
" Regius Professor," was not a member) until January, 1830, when his
efforts were crowned with success. A plot of ground was then pur-
chased at the corner of College Street and Shuttle Street, outside the
College precincts, and operations were at once begun, and pushed
on with such vigour that the buildings seem to have been finished
towards the end of the same year. The building thus erected contained ample and well-designed accommodation for teaching and
experimental work. There was a large class-room and a large and
conveniently arranged public laboratory for students, with private
rooms for the professor and for the prosecution of experimental
research by the professor and his assistants, or by students and
others.

Part of the ground floor of the premises was let to a tenant (the
" Falstaff Tavern " for many years !). To-day I found the building
still in existence, and occupied by " George Younger and Co.'s Yarn
Stores." Nearly all the rest of the University Buildings within the
College precincts have been pulled down within the last twelve
years for the " College Railway Station," which now occupies the
site of the old Glasgow College and University.—W. T.]

was attended by a large number of students. These were chiefly medical students, but a considerable number also were students who wished to learn chemistry to practise it in the various chemical manufactories in Glasgow and the North of England, while some went to learn chemistry solely for the sake of science. A chemical laboratory has now become indispensable in all universities. A notable development of chemical laboratories with reference to practical education in chemistry, was made by Liebig not many years after 1831. I fix that date from personal recollection. In 1831 I first came to Glasgow, and I well remember that the building containing the chemical lecture room and laboratory existed then. How long before 1831 it was built I do not at this moment recollect. The world-renowned laboratory of Liebig brought together all the young chemists of the day. If I were to name the great men who studied at Giessen I should have to name almost every one of the great chemists of the present day who were young forty years ago. His laboratory was in full and flourishing activity between 1841 and 1845, and continued so for several years more until he migrated to Munich. It is still, I believe, a prosperous institution, carrying out the aims of its founder with undiminished zeal and energy. One of those chemists now living, who was young forty years ago, told me a few days since that Liebig's laboratory looked like an old stable. I believe the building in which we are now assembled *was* an old stable, but I fail to discover that it looks like an old stable now. If Liebig's laboratory, looking like an old stable, brought out such results to astonish and benefit the world, what must we expect of the beautiful laboratory in which we are now met? What would Liebig not have given for the appliances and advantages afforded by the well-equipped buildings of the North Wales College at Bangor? What would Liebig not have given for the facilities which now exist in these admirably-appointed lecture-rooms in which we are now met, and for the carefully-equipped laboratories and working-rooms, and places for special experimental work covering the

area of the old stables and coach-houses of the
" Penrhyn Arms Hotel"! If the professors and
the students in this College—I think I may
already say this thriving College—will be inspired
by the zeal of those who have worked before
them, a great reward will result even in the first
year of the existence of the institution.

With respect to physical laboratories I may be
allowed, without being thought egotistical, to say
something in which I must speak of my own
action. The physical laboratory in the University
of Glasgow is, I believe, the first of the physical
laboratories of which we have now so many
When I entered upon the professorship of natural
philosophy at Glasgow I found apparatus of a very
old-fashioned kind. Much of it was more than a
hundred years old, little of it less than fifty years
old, and most of it was of worm-eaten mahogany.
Still with such appliances year after year students
of natural philosophy had been brought together
and taught. The principles of dynamics and
electricity had been well illustrated and well
taught : as well taught as lectures and so imperfect

apparatus—but apparatus merely of the lecture-illustration kind—could teach. But there was absolutely no provision of any kind for experimental investigation, still less idea, even, for anything like students' practical work. Students' laboratories in physical science were not then thought of. I remember one of the chemists of the Liebig school asking me what was the object of a physical laboratory. I replied that it was to investigate the properties of matter. I could give no better answer now. I may remind you that there is no philosophical division whatever between chemistry and physics. The distinction is that different properties are investigated by different sets of apparatus. The distinction between chemistry and physics must be merely a distinction of detail and of division of labour.

Soon after I entered my present chair in the University of Glasgow in 1846 I had occasion to undertake some investigations of certain electro-dynamic qualities of matter, to answer questions which had been suggested by the results of mathematical theory, questions which could only be answered by direct experiment. The labour of observing proved too heavy, much of it could scarcely be carried on without two or more persons working together. I therefore invited students to aid in the work. They willingly accepted the invitation, and lent me most cheerful and able help. Soon after, other students, hearing that some of their class-fellows had got experimental work to do, came to me and volunteered to assist in the investigation. I could not give them all work in the particular investigation with which I had commenced—"The electric convection of heat"—for want of means and time and possibilities of arrangement, but I did all in my power to find work for them on allied subjects (Electrodynamic Properties of Metals,[1] Moduluses of Elasticity of Metals, Elastic Fatigue, Atmospheric Electricity, &c.) I then had an ordinary class of a hundred students, of whom some attended lectures in natural philosophy two hours a day, and

---

[1] Results up to 1856, published under this title, as Bakerian Lecture for 1856 (*Trans. R. S.*, and republished recently in vol. ii. of *Collected Papers.*—W. T.

had nothing more to do from morning till night. Those were the palmy days of natural philosophy in the University of Glasgow—the pre-Commissional days. But the majority of the class really had very hard work, and many of them worked after class-hours for self-support. Some were engaged in teaching, some were city-missionaries, intending to go into the Established Church of Scotland or some other religious denomination of Scotland, or some of the denominations of Wales, for I always had many Welsh students. But about five and twenty of the whole number found time to come to me for experimental work several hours every day. In those days, as now, in the Scottish Universities all intending theological students took the "philosophical curriculum"—*zuerst collegium logicum*—then moral philosophy, and (generally last) natural philosophy. Three-fourths of my volunteer experimentalists used to be students who entered the theological classes immediately after the completion of the philosophical curriculum. I well remember the surprise of a great German Professor when he heard of this rule and usage :

" What ! do the theologians learn physics ? " I
said, " Yes, they all do ; and many of them have
made capital experiments." I believe they do not
find that their theology suffers at all from having
learned something of mathematics, and dynamics
and experimental physics before they enter upon
it. I had then no other premises than the old
lecture-room and the adjoining apparatus room.
To meet my requirements for my new volunteer
laboratory corps, the " Faculty (the then governing body of the College) allotted to me an old wine-
cellar, part of an old professor's house, the rest of
which had been converted into lecture-rooms.
This, with the bins swept away, and a water-
supply and a sink added, served as physical
laboratory (a name then unknown) for several
years, till the University Commissioners came and
abolished a certain old function of Glasgow
University, the " Blackstone Examination." The
examination room was left unprotected, its talis-
man, the old " Blackstone Chair," removed. I
instantly annexed it (it was very convenient, adjoining the old wine-cellar and below the apparatus

room) ; and, as soon as it could conveniently be done, obtained the sanction of the Faculty for the annexation. The Blackstone room and the old wine-cellar served well for physical laboratory till 1870, when the University was removed from its old site imbedded in the densest part of the city, to the airy hill-top on which it now stands. In the new University buildings ample and commodious provision was made for experimental work.

In that good old time some students used to come to me under the impression that the laboratory would prove an agreeable lounge, where they could meet pleasantly and spend the forenoon talking matters over. They were soon undeceived as to its being a lounge for idly whiling away time. I hope they were not altogether disappointed when they thought it would be agreable, and I almost hope they found it even more agreeable than they expected. They certainly learned something of patience and perseverance, if not much science, in the six months of the College session. As a matter of general education for those not going to practise medicine, was it of any use entering a chemical or physical laboratory ? I found as many as three-quarters of the students were destined for service in the religious denominations in after-life. I have frequently met some of those old students who had entered upon their profession as ministers, and have found that they always recollected with interest their experimental work at the University. They felt that the time they had spent in making definite and accurate measurements had not been time thrown away, because it educated them into accuracy,—it educated them into perseverance if they required such education. Some students even worked so hard in my laboratory that I had to interpose for the sake of their health. There is one thing I feel strongly in respect to investigation in physical or chemical laboratories—it leaves no room for shady, doubtful distinctions between truth, half-truth, whole falsehood. In the laboratory everything tested or tried is found either true or not true. Every result is *true.* Nothing not proved true is a *result ;*—there is no such thing as doubtfulness. The search for absolute and unmistakable truth is promoted by laboratory work in a manner beyond all conception. It is a kind of work in which also patience and perseverance are promoted in a most marked degree. No labour must be shrunk from ; everything must be carefully done. There is this which is satisfactory about it : that perseverance is sure to be rewarded. There is no failure in physical science. We do not always find the particular thing looked for ; we often find that what we looked for does not exist, or that something else exists very different from what we expected to find ; but that something is to be found in any investigation entered upon with intelligence and pursued with perseverance, is a certainty ; and also that that something is not valueless follows as a matter of course. Every additional knowledge of the properties of matter is of value.

A large part of the work of a physical or chemical laboratory must be measurement. That might seem rather trying work ; "harsh and crabbed" shall we say ? Who cares to measure the length of a line in land-surveying, or of a piece of cord, or of ribbon, or of cloth ? These may not be in themselves essentially interesting occupations; but if it becomes necessary to measure something smaller than can be seen with the eye, the measurement itself becomes an object to inspire the worker with interested ardour. Dulness does not exist in science. What do you think of a measurement of something you can only gauge by inference from the performance of the apparatus tested in some peculiarly subtle way? The difficulties to be overcome in physical science in mere measurement are teeming with interest. Properties of matter, or forces to be contended with, oblige us to be always digressing. We cannot go on saying—"We will think of nothing but the object before us." Every person who aims at one object of course perseveres until he attains it ; but he keeps his mind open until he can return to some other object never thought of at first, but which thrust itself on him as a difficulty occurring in the pursuit of the first object. The very disappointments in attaining objects sought after in the investigations of physical science are the richest sources of ultimate profit, and present satisfaction and pleasure, notwithstanding the difficulties and disappointments contended with. But I am afraid I am taxing your patience too much. I will only just say with reference to physical laboratories that they are now advancing to something of the method and consistent system that Thomas Thomson and Liebig so greatly gave to chemical laboratories. I, myself, have not done so much as I might have done in that way. The physical laboratory at Glasgow has, I believe, been, more than most others, devoted to whatever work occurred in physical investigation, measuring properties of matter, comparing thermometers, electrometers, galvanometers, and doing other practically useful work. We put the junior students at once into investigations, and let them measure and weigh whatever requires measurement and weighing in the course of the investigation. I look with admiration to what has been done by those who have worked up physical laboratories to their present advanced condition. The physical laboratories of King's College and University College,

London, under the admirable organisation and work of Professor Adams and Professor Carey Foster ; the Cavendish laboratory at Cambridge, originated by Clerk Maxwell, and admirably systematised and perfected by Lord Rayleigh, have rendered splendid services to physical science all over the world. Much has been done even to provide suitable text-books for use in the systematic practical training of students in laboratory work : for example, the *Treatise on Physical Measurement* by Kohlrausch, which has been for several years a most serviceable manual, and the lately published *Practical Physics* of Glazebrook and Shaw. The physical laboratory system has now become quite universal. No university in the world can now live unless it has a well-equipped laboratory. I hope you will all do your best to make the physical and chemical laboratories of this College a great success ; that you will follow example in everything exemplary until the Bangor laboratories become a model to be followed in future laboratories in Wales, England, or any other part of the world, I was not quite accurate when I spoke of this new college in this City of Bangor as *the* University College of North Wales. My friend, Mr. Cadwaladr Davies, your secretary, has reminded me that there was a university of North Wales at Bangor-is-y-coed, in Flintshire—not a city, because it did not combine a bishop and a mayor—but a town which had the honour of having been the seat of the first Welsh university known to history. There may have been universities in Wales before the one which flourished 1200 years ago at Bangor-is-y-coed ; but their history is lost in the long night of silence, because no sacred bard sung of their existence. The university of Bangor-is-y-coed had its bard, who tells us that the institution had 2100 students. There you have a worthy object of ambition for the city of Bangor ! May it soon have a goodly proportion of the 2100. Perhaps not so long a time may elapse before your college and the other colleges in Wales may reach to such a number. Indeed, I do not see anything un-reasonable in hoping and expecting that in a dozen years there will be 2100 university-students in Wales. The population of Wales is more than a million and a half, which is, I think, about a fourth of the population of Scotland ; and I do not see why Wales should not have university students in proportion to its population as well as Scotland. I believe the brightness and activity of the Welsh intelligence will thoroughly take up the idea of a university, and profit by it to the utmost, and, I believe, the existence of this institution at Bangor will before twenty years have passed away be looked upon as having been a great benefit to the Principality. What Wales gained by the university at Bangor-is-y-coed can scarcely now be told, but alas, for that university with its 2000 students, it was destroyed in the year 613 by Ethelfred, King of Northumbria, and its destruction was followed by 900 years of dark ages. Thus we see what the world lost by the annihilation of the first university of North Wales. Another bard, Lewis Glencothy, advocated and sang of the possibility of a university in Wales in the time of Henry VII. Richard Baxter, not a Welshman nor a bard but the great English Puritan divine

reported to the then Government under Cromwell
in favour of a university for Wales. Cromwell
died before action was taken, and nothing was
done in the matter for nearly 200 years, when a
very active desire sprang up and active co-operation among all parties was entered upon, for
having a university established in Wales. We see
everything now prospering in that direction. I
look forward hopefully to the time when this
college of Bangor—if not an independent university of its own—will be a college of the
University of Wales. All the colleges of Wales,
equipped to do the work of a university, might be
united to form a University of Wales. There are
very many important advantages in favour of such
an arrangement. No doubt it is an object of
honourable ambition ; but it may be asked if a
college does all the work of a university, what
does it matter whether it is called a university or
not ? It is of considerable importance that your
college should be either a university itself or part
of a university of which it is an integral college.
One of the advantages would be that the teaching

of the college would be enabled to take a more practical form than it can possibly take as long as its main purpose is that of preparing students for the degree examinations of London University. The degree system of London University fills a widespread want—a want felt over the whole range of the British empire ; a want of marking by the stamp of a university degree, if not by some more suitable title, the possession of knowledge and of a certain amount of training by those who have not had the opportunity of obtaining that knowledge in any thoroughly equipped college or university. That is a splendid reason for the existence of the London University, and it has well fulfilled its reason for existence. But, for all that, it would be greatly better for the students of the University College of North Wales if the teaching were conducted with reference to an examination carried on by their own professors and colleague professors in other properly equipped Welsh colleges. It is the greatest mistake in respect to teaching and examining to think that the examiner is an inspector. An examiner of schools must to some extent take that position. But in university work teaching and examining must go side by side, hand in hand, day by day, week by week, together, if the work is to be well done. The object of a university is teaching, not testing. Testing products comes at some times, and for some special purposes, to be a necessity; but in respect to the teaching of a university, the object of examination is to promote the teaching. The examination should be, in the first place, daily. No professor should meet his class without talking to them. He should talk to them and they to him. The French call a lecture a *conférence*, and I admire the idea involved in that name. Every lecture should be a conference of teacher and students. It is the true ideal of a professorial lecture. I have found that many students are afflicted when they come up to college with the disease called "aphasia." They will not answer when questioned, even when the very words of the answer are put in their mouths, or when the answer is simply "yes" or "no." That disease wears off in a few weeks, but the great cure for it is in repeated and careful and very free interchange of question and answer between teacher and student. Professors and students must speak to one another. One of the greatest things is to promote freedom of conversation in such classes, to cultivate in them the power of expressing ideas in words. Then something more definite than *vivâ voce* examination can come. Written examinations are very important, as training the student to express with clearness and accuracy the knowledge he has gained, and to work out problems, or numerical results, but they should be once a week to be beneficial. If only occurring once in two or three months they will lose their effect in promoting good teaching, and can be scarcely more than a test ; if only once a year they are merely inspector's work. The object of the university should be teaching, and examining should only be part of its work, and that only so far as it promotes teaching. The credit of the university should depend on good teaching, and no candidate should be granted a degree who does not show that he has taken advantage of the good

K K 2

teaching. But it is impossible to carry out that
programme to best advantage by a college which
is not in itself an integral part of a university.
Such examinations as those of the London University are necessarily arranged to suit thousands
of candidates who have learned in different schools,
and cannot always contain questions that would
be most suitable for one particular mode of
teaching The kind of questions set would be of
a different nature if the giving of the questions
devolved upon those who had in hand the teaching.
Those who have the teaching can give an examination vastly more useful and one that would
re-act on the teaching in a way that an examination of a multitude of students trained at all kinds
of institutions, and many merely by private
reading, could not possibly do. Therefore, it
seems to be a matter of high importance indeed
that there should be a University of Wales ; that
you should consider it to be a great object to be
attained, sooner or later—but the sooner the
better—the establishment of the University of
Wales, with the University College of North

Wales an integral part of it. I have much pleasure in wishing the University College of North Wales every success, and I trust that the laboratories now opened may prove of great value in promoting and aiding the study of science.

PRESIDENTIAL ADDRESSES.

[*Delivered at the Anniversary Meetings of the Royal Society, of November, 30th, 1891, November 30th, 1892, and November 30th, 1893.*]

EXTRACT FROM ADDRESS OF NOVEMBER 30TH, 1891.

A FUNDAMENTAL investigation in astronomy, of great importance in respect to the primary observational work of astronomical observatories, and of exceeding interest in connection with tidal, meteorological, and geological observations and speculations, has been definitely entered upon during the past year, and has already given substantial results of a most promising character. The International Geodetic Union, at its last meeting in the autumn of 1890, on the motion of Professor Foerster, of Berlin, resolved to send an astronomical expedition to Honolulu, which is within 9° of the opposite meridian to Berlin (171° west from Berlin), for the purpose of making a twelve months' series of observations on latitude corresponding to twelve months' analogous observations to be made in the Royal Observatory, Berlin. Accordingly Dr. Marcuse went from Berlin, and, along with Mr. Preston, sent by the Coast and Geodetic Survey Department of the United States, began making latitude observations in Honolulu about the beginning of June. In a letter from Professor Foerster, received a few weeks ago, he tells me that he has already received from Honolulu a first instalment of several hundred determinations of latitude, made during a first three months of the proposed year of observations; and that, in comparing these results with the corresponding results of the Berlin Observatory, he finds beyond doubt that in these three months the latitude increased in Berlin by one-third of a second and decreased in Honolulu by almost exactly the same amount. Thus, we have decisive demonstration that motion, relatively to the Earth, of the Earth's instantaneous axis of rotation, is the cause of variations of latitude which had been observed in Berlin, Greenwich, and other great observatories, and which could not be wholly attributed to errors of observation. This, Professor Foerster remarks, gives observational proof of a dynamical conclusion contained in my Presidential Address to Section A of the British Association, at Glasgow, in 1876, to the effect that irregular movements of the Earth's axis to the extent of half a second may be produced by the temporary changes of sea-level due to meteorological causes.

It is proposed that four permanent stations for regular and continued observation of latitude, at places of approximately equal latitude and on meridians approximately 90° apart, should be established under the auspices of the International Geodetic Union. The reason for this is that a change in the instantaneous axis of rotation in the direction perpendicular to the meridian of any one place would not alter its latitude, but would alter the latitude of a place 90° from it in longitude by an amount equal to the angular change of the position of the axis. Thus two stations in meridians differing by 90° would theoretically suffice, by observations of latitude, to determine the changes in the position of the instantaneous axis ; but differential results, such as those already obtained between Berlin and Honolulu, differing by approximately 180° in longitude, are necessary for eliminating errors of observation sufficiently to give satisfactory and useful results. It is to be hoped that England, and all other great nations in which science is cultivated, will co-operate with the International Geodetic Union in this important work.

Among the most interesting scientific events of the past year was the celebration of the hundredth anniversary of the birth of Faraday by the two Faraday Lectures in the Royal Institution last June. In the first of these, which was delivered by Lord Rayleigh, under the presidency of the Prince of Wales, an old pupil of Faraday's and now Vice-Patron of the Royal Institution, a general survey of Faraday's work during his fifty-four years' connexion with the Royal Institution was given. Naturally, a large part of the lecture was devoted to magnetism and electricity and to electro-magnetic induction; but it contained also much that must have been surprising to the audience, scarcely prepared to be told, as they were told by Lord Rayleigh, that "Faraday's mind was essentially mathematical in its qualities," and that, particularly in his acoustical work, he had made many very acute observations of physical phenomena, of a kind to help in guiding the mathematician to the solution of difficult and highly interesting problems of mathematical dynamics, and in some cases actually to give him the solution surprisingly different from what might have been expected even by highly qualified mathematical investigators.

The other Faraday Lecture, given by Professor Dewar, was a splendid realisation of Faraday's anticipations regarding the liquefaction of the "permanent gases," according to which no ex-

treme of pressure might be capable of liquefying
hydrogen or oxygen at ordinary temperature,
while a very moderate pressure might suffice to
liquefy them if their temperatures could be sufficiently lowered. Professor Dewar actually showed
liquid oxygen in a glass tumbler, not boiling or
in a state of commotion like a tumbler of soda-
water, but quietly and without any sensible motion
keeping itself cool by its own evaporation, while
it rapidly formed a thick jacket of hoar-frost on
the outside of the vessel by condensation of
watery vapour from the surrounding atmosphere.
The surprise and delight of the audience reached
a climax when liquid oxygen was poured from one
open vessel to another before their eyes.

A matter of great importance in respect to the
health of the community was submitted to the
Royal Society by the London County Council, in
a letter of date May 1, 1891, asking for information and suggesting investigation regarding the
vitality of microscopic pathogenic organisms in
large bodies of water, such as rivers which are
sources of water-supply and which are exposed to

contamination. After some correspondence it was agreed, between the County Council and the Council of the Royal Society, to enter upon an investigation, the expense of which was to be defrayed partly by the London County Council and partly by the Royal Society out of the Government Grant for Scientific Research. When we consider how much of disease and death is due to contaminated water, we must feel that it is scarcely possible to overestimate the vital importance of the proposed investigation. Let us hope that the alliance between the London County Council and the Royal Society, for this great work, may be successful in bringing out practically useful results.

Extract from Address of November 30th, 1892.

Mr. Elis's communication [1] to the Royal Society of last May, and Professor Grylls Adams' communication [2] of June, 1891, both on the subject of

---

[1] *Roy. Soc. Proc.*, November, 1822, vol. 52, p. 191.
[2] *Phil. Trans.*, vol. 183, 1891–92, p. 131.

simultaneous magnetic disturbances found by observations at magnetic observatories in different parts of the world ; the award of a Royal medal two years ago to Hertz, for his splendid experimental work on electro-magnetic waves and vibrations ; and Professor Schuster's communication [1] to the Royal Society, of June, 1889, on the " Diurnal Variations of Terrestrial Magnetism " ; justify me in saying a few words on the present occasion regarding terrestrial magnetic storms, and the hypothesis that they are due to magnetic waves emanating from the sun.

Guided by Maxwell's "electro-magnetic theory of light," and the undulatory theory of propagation of magnetic force which it includes, we might hope to perfectly overcome a fifty years' outstanding difficulty in the way of believing the sun to be the direct cause of magnetic storms in the earth, though hitherto every effort in this direction has been disappointing. This difficulty is clearly stated by Professor W. G. Adams, in the following sentences, which I quote from his Report to the

[1] *Roy. Soc. Proc.*, vol, 180, 1892, p. 467.

British Association of 1881 (p. 469), "On Magnetic Disturbances and Earth Currents":—"Thus we see that the magnetic changes which take place at various points of the earth's surface at the same instant are so large as to be quite comparable with the earth's total magnetic force; and in order that any cause may be a true and sufficient one, it must be capable of producing these changes rapidly."

The primary difficulty, in fact, is to imagine the sun a variable magnet or electro-magnet, powerful enough to produce at the earth's distance changes of magnetic force amounting, in extreme cases, to as much as $\frac{1}{20}$ or $\frac{1}{30}$, and frequently, in ordinary magnetic storms, to as much as $\frac{1}{100}$ of the undisturbed terrestrial magnetic force.

The earth's distance from the sun is 228 times the sun's radius, and the cube of this number is about 12,000,000. Hence, if the sun were, as Gilbert found the earth to be, a globular magnet, and if it were of the same average intensity of magnetisation as the earth, we see, according to the known law of magnetic force at a distance, that the magnetic force due to the sun at the earth's distance from it, in any direction, would be only a twelve-millionth of the actual force of terrestrial magnetisation at any point of the earth's surface, in a corresponding position relatively to the magnetic axis. Hence the sun must be a magnet [1] of not much short of 12,000 times the average intensity of the terrestrial magnet (a not absolutely inconceivable supposition, as we shall presently see) to produce, by direct action simply as a magnet, any disturbance of terrestrial magnetic force sensible to the instruments of our magnetic observatories.

Considering probabilities and possibilities as to the history of the earth from its beginning to the present time, I find it unimaginable but that terrestrial magnetism is due to the greatness and the rotation of the earth. If it is true that

---

[1] The moon's apparent diameter being always nearly the same as the sun's, the statements of the last four sentences are applicable to the moon as well as to the sun, and are important in connection with speculation as to the cause of the lunar disturbance of terrestrial magnetism, discovered nearly fifty years ago by Kreil and Sabine.

terrestrial magnetism is a necessary consequence of the magnitude and the rotation of the earth, other bodies comparable in these qualities with the earth, and comparable also with the earth in respect to material and temperature, such as Venus and Mars, must be magnets comparable in strength with the terrestrial magnet, and they must have poles similar to the earth's north and south poles on the north and south sides of their equators, because their directions of rotation, as seen from the north side of the ecliptic, are the same as that of the earth. It seems probable, also, that the sun, because of its great mass and its rotation in the same direction as the earth's rotation, is a magnet with polarities on the north and south sides of its equator, similar to the terrestrial northern and southern magnetic polarities. As the sun's equatorial surface-velocity is nearly four and a half times the earth's, it seems probable that the average solar magnetic moment exceeds the terrestrial considerably more than according to the proportion of bulk. Absolutely ignorant as we are regarding the effect of cold

solid rotating bodies such as the earth, or
Mars, or Venus, or of hot fluid rotating bodies
such as the sun, in straining the circumambient
ether, we cannot say that the sun might not be
1000, or 10,000, or 100,000 times as intense a
magnet as the earth. It is, therefore, a perfectly
proper object for investigation to find whether
there is, or is not, any disturbance of terrestrial
magnetism, such as might be produced by a
constant magnet in the sun's place with its magnetic axis coincident with the sun's axis of rotation. Neglecting for the present the seven degrees
of obliquity of the sun's equator, and supposing
the axis to be exactly perpendicular to the ecliptic,
we have an exceedingly simple case of magnetic
action to be considered : a magnetic force perpendicular to the ecliptic at every part of the earth's
orbit and varying inversely as the cube of the
earth's distance from the sun. The components of
this force parallel and perpendicular to the earth's
axis are, respectively, 0·92 and 0·4 of the whole ;
of which the former could only be perceived in
virtue of the varying distance of the earth from the

sun in the course of a year; while the latter would give rise to a daily variation, the same as would be observed if the red ends of terrestrial magnetic needles were attracted towards an ideal star of declination 0° and right ascension 270°. Hence, to discover the disturbances of terrestrial magnetism, if any there are, which are due to the direct action of the sun as a magnet, the photographic curves of the three magnetic elements given by each observatory should be analysed for the simple harmonic constituent of annual period and the simple harmonic constituent of period equal to the sidereal day. We thus have two very simple problems, each of which may be treated with great ease separately by a much simplified application of the principles on which Schuster has treated his much more complex subject, according to Gauss' theory as to the external or internal origin of the disturbance, and Professor Horace Lamb's investigation of electric currents induced in the interior of a globe by a varying external magnet. The sidereal diurnal constituent which forms the subject of the second of these simplified problems is smaller, but not much smaller, than the solar diurnal term which, with the solar semi-diurnal, the solar ter-diurnal, and solar quarter-diurnal constituents, form the subjects of Schuster's paper. The conclusion at which he has arrived, that the source of the disturbance is external, is surely an ample reward for the great labour he has bestowed on the investigation hitherto; and I hope he may be induced to undertake the comparatively slight extension of his work which will be required for the separate treatment of the two problems of the sidereal diurnal and the solar annual constituents, and to answer for each the question :—Is the source external or internal?

But even though external be the answer found in each case, we must not from this alone assume that the cause is direct action of the sun as a magnet. The largeness of the solar semi-diurnal, ter-diurnal, and quarter-diurnal constituents found by the harmonic analysis, none of which could be explained by the direct action of the sun as a magnet, demonstrate relatively large action of some other external influence, possibly the electric currents in our atmosphere, which Schuster suggested as a probable cause. The cause, whatever it may be, for the semidiurnal and higher constituents would also probably have a variation in the solar diurnal period on account of the difference of temperature of night and day, and a sidereal and annual period on account of the difference of temperature between winter and summer.

Even if, what does not seem very probable, we are to be led by the analysis to believe that magnetic force of the sun is directly perceptible here on the earth, we are quite certain that this steady force is vastly less in amount than the abruptly varying force which, from the time of my ancestor in the Presidential Chair, Sir Edward Sabine's discovery,[1] forty years ago, of an apparent connexion between sun-spots and terrestrial magnetic storms, we have been almost compelled to attribute to disturbing action of some kind at the sun's surface.

[1] Communication to the Royal Society, March 18th, 1852 (*Phil. Trans.*, vol. 162, p. 143).

As one of the first evidences of this belief, I may quote the following remarkable sentences from Lord Armstrong's Presidential Address to the British Association at Newcastle, in 1863 :—

" The sympathy also which appears to exist between forces operating in the sun and magnetic forces belonging to the earth merits a continuance of that close attention which it has already received from the British Association, and of labours such as General Sabine has, with so much ability and effect, devoted to the elucidation of the subject. I may here notice that most remarkable phenomenon which was seen by independent observers at two different places, on the 1st of September, 1859. A sudden outburst ot light, far exceeding the brightness of the sun's surface, was seen to take place, and sweep like a drifting cloud over a portion of the solar face. This was attended with magnetic disturbances of unusual intensity, and with exhibitions of aurora of extraordinary brilliancy. The identical instant at which the effusion of light was observed was recorded by an abrupt and strongly marked deflection in the self-registering instruments at Kew. The phenomenon as seen was probably only part of what actually took place, for the magnetic storm in the midst of which it occurred commenced before, and continued after, the event. If conjecture be allowable in such a case, we may suppose that this remarkable event had some connexion with the means by which the sun's heat is renovated. It is a reasonable supposition that the sun was at that time in the act of receiving a more than usual accession of new energy; and the theory which assigns the maintenance of its power to cosmical matter, plunging into it with that prodigious velocity which gravitation would impress upon it as it approached to actual contact with the solar orb, would afford an explanation of this sudden exhibition of intensified light, in harmony with the knowledge we have now attained, that arrested motion is represented by equivalent heat.''

It has certainly been a very tempting hypothesis, that quantities of meteoric matter suddenly falling into the sun is the cause, or one of the causes, of those disturbances to which magnetic storms on the earth are due. We may, indeed, knowing that meteorites do fall into the earth, assume without doubt that much more of them fall, in the same time, into the sun. Astronomical reasons, however, led me long ago to conclude that their quantity annually, or per century, or per thousand years, is much too small to supply the energy given out by the sun in heat and light radiated through space, and led me to adopt unqualifiedly Helmholtz's theory, that work done by gravitation on the shrinking mass is the true source of the sun's heat, as given out at present, and has been so for several hundred thousand years, or several million years. It is just possible, however, that the outburst of brightness described by Lord Armstrong may have been due to an extraordinarily great and sudden falling in of meteoric matter, whether direct from extra-planetary space or from orbital circulation round the sun. But it seems to me much more probable that it was due to a refreshed brightness produced over a larger area of the surface than usual by brilliantly incandescent fluid rushing up from below, to take the place of matter falling down from the surface, in consequence of being cooled in the regular *régime* of solar radiation. It seems, indeed, very improbable that meteors fall in at any time to the sun in sufficient quantity to produce dynamical disturbances at his surface at all comparable with the gigantic storms actually produced by hot fluid rushing up from below, and spreading out over the sun's surface.

But now let us consider for a moment the work which must be done at the sun to produce a terrestrial magnetic storm. Take, for example, the magnetic storm of June 25, 1885, of which Adams gives particulars in his paper of June, 1891 (*Phil. Trans.*, p. 139 and Pl. 9). We find at eleven places, St. Petersburg, Stonyhurst, Wilhelmshaven, Utrecht, Kew, Vienna, Lisbon, San Fernando, Colaba, Batavia, and Melbourne, the horizontal force increased largely from 2 to 2.10 P.M., and fell at all the places from 2.10 to 3 P.M., with some rough ups and downs in the interval. The storm lasted altogether from about noon to 8 P.M. At St. Petersburg, Stonyhurst, and Wil-

helmshaven, the horizontal force was above par by
0·00075, 0·00088, and 0·00090 (C.G.S. in each case)
at 2.10 P.M. ; and below par by 0·0007, 0·00066,
0·00075 at 3 o'clock. The mean value for all the
eleven places was nearly 0·0005 above par at 2h.
10m., and 0·0005 below par at 3h. The photographic curves show changes of somewhat similar
amounts following one another very irregularly,
but with perfectly simultaneous correspondence at
the eleven different stations, through the whole
eight hours of the storm. To produce such changes
as these by any possible dynamical action within
the sun, or in his atmosphere, the agent must have
worked at something like 160 million million
million million horse-power[1] (12×10^{35} ergs per
sec.), which is about 364 times the total horse-
power ($3·3 \times 10^{33}$ ergs per sec.) of the solar radiation. Thus, in this eight hours of a not very
severe magnetic storm, as much work must have
been done by the sun in sending magnetic waves
out in all directions through space as he actually
does in four months of his regular heat and light.

[1] 1 horse power $= 7·46 \times 10^9$ ergs per second.

This result, it seems to me, is absolutely conclusive against the supposition that terrestrial magnetic storms are due to magnetic action of the sun ; or to any kind of dynamical action taking place within the sun, or in connexion with hurricanes in his atmosphere, or anywhere near the sun outside.

It seems as if we may also be forced to conclude that the supposed connexion between magnetic storms and sun-spots is unreal, and that the seeming agreement between the periods has been a mere coincidence.

We are certainly far from having any reasonable explanation of any of the magnetic phenomena of the earth; whether the fact that the earth is a magnet ; that its magnetism changes vastly, as it does from century to century ; that it has somewhat regular and periodic annual, solar diurnal, lunar diurnal, and sidereal diurnal variations ; and (as marvellous as the secular variation) that it is subject to magnetic storms. The more marvellous, and, for the present, inexplicable, all these subjects are, the more exciting becomes the pursuit of investigations which must, sooner or later, reward those who persevere in the work. We have at present two good and sure connexions between magnetic storms and other phenomena : the aurora above, and the earth currents below, are certainly in full working sympathy with magnetic storms. In this respect the latter part of Mr. Ellis's paper is of special interest, and it is to be hoped that the Greenwich observations of earth currents will be brought thoroughly into relation with the theory of Schuster and Lamb, extended, as indeed Professor Schuster promised to extend it, to include not merely the periodic diurnal variations, but the irregular sudden changes of magnetic force taking place within any short time of a magnetic storm.

In my Presidential Address of last year I referred to the action of the International Geodetic Union, on the motion of Professor Foerster, of Berlin, to send an astronomical expedition to Honolulu for the purpose of making a twelve months' series of observations on latitude corresponding to twelve months' simultaneous observations to be made in European observatories ; and I was enabled, through the kindness of Professor Foerster, to announce as a preliminary result, derived from the first three months of the observations, that the latitude had increased during that time by $\frac{1}{8}$ sec. at Berlin, and had decreased at Honolulu by almost exactly the same amount. The proposed year's observations, begun in Honolulu on the 1st of June, 1891, were completed by Dr. Marcuse, and an elaborate reduction of them by the permanent Committee of the International Geodetic Union was published a month ago at Berlin. The results are in splendid agreement with those of the European observatories: Berlin, Prag, and Strasbourg. They prove beyond all question that between May, 1891, and June, 1892, the latitude of each of the three European observatories was a maximum, and of Honolulu a minimum, in the beginning of October, 1891: that the latitude of the European observatories was a minimum, and of Honolulu a maximum, near the beginning of May, 1892: and that the variations during the year followed somewhat approximately, simple harmonic law as if for a period of 385 days, with range of about $\frac{1}{4}$ sec. above and below the mean latitude in each case. This is just what would result from motion of the north and south polar ends of the earth's instantaneous axis of rotation, in circles on the earth's surface of 7˙5 metres radius, at the rate of once round in 385 days.

Some time previously it had been found by Mr. S. C. Chandler that the irregular variations of latitude which had been discovered in different observatories during the last fifteen years seemed to follow a period of about 427 days, instead of the 306 days given by Peters' and Maxwell's dynamical theory, on the supposition of the earth being wholly a rigid body. And now, the German observations, although not giving so long a period as Chandler's, quite confirm the result that, whatever approximation to following a period there is, in the variations of latitude, it is a period largely exceeding the old estimate of 306 days.

Newcomb, in a letter which I received from him last December, gave, what seems to me to be, undoubtedly, the true explanation of this apparent discrepance from dynamical theory, attributing it to elastic yielding of the earth as a whole. He added a suggestion specially interesting to myself, that investigation of periodic variations of latitude may prove to be the best means of determining approximately the rigidity of the earth. As it is we have now, for the first time, what seems to be a quite decisive demonstration of elastic yielding in the earth as a whole, under the influence of a deforming force, whether of centrifugal force round a varying axis, as in the present case, or of tide-generating influences of the sun and moon with reference to which I first raised the question of elastic yielding of the earth's material many years ago.

The present year's great advance in geological dynamics forms the subject of a contribution by Newcomb to the *Monthly Notices of the Royal Astronomical Society*, of last March. In a later paper, published in the *Astronomische Nachrichten*, he examines records of many observatories, both of Europe and America, from 1865 to the present time, and finds decisive evidence that from 1865 to 1890 the variations of latitude were much less than they have been during the past year, and seeming to show that an augmentation took place, somewhat suddenly, about the year 1890.

When we consider how much water falls on Europe and Asia during a month or two of rainy season, and how many weeks or months must pass before it gets to the sea, and where it has been in the interval, and what has become of the air from which it fell, we need not wonder that the distance of the earth's axis of equilibrium of centrifugal force from the instantaneous axis of rotation should often vary[1] by five or ten metres in the course of a few weeks or months. We can scarcely expect, indeed, that the variation found by the International Geodetic Union during the year beginning June, 1891, should recur periodically for even as much as one or two or three times of the seeming period of 385 days.

One of the most important scientific events of the past year has been Barnard's discovery, on the 9th of September, of a new satellite to Jupiter. On account of the extreme faintness of the object it has not been observed anywhere except at the

[1] See *Brit. Assoc. Reports*, 1876, Address to Section A, pp. 10, 11.

Lick Observatory in California. There, at an elevation of 4,500 ft., with an atmosphere of great purity, and with a superb refractor of 36″ aperture, they have advantages not obtainable elsewhere. The new satellite is about 112,000 miles distant from Jupiter, and its periodic time is about 11 h. 50 m. Mr. Barnard concludes a short statement of his discovery with the following sentences :—
" It will thus be seen that this new satellite makes two revolutions in one day, and that its periodic time about the planet is less than two hours longer than the axial rotation of Jupiter. Excepting the inner satellite of Mars, it is the most rapidly revolving satellite known. When sufficient, observations have been obtained, it will afford a new and independent determination of the mass of Jupiter. Of course, from what I have said in reference to the difficulty of seeing the new satellite, it will be apparent that the most powerful telescopes of the world only will show it " (dated Mount Hamilton, September 21, 1892).

Sir Robert Ball, in calling my attention to it remarks that " it is by far the most striking addition to the solar system since the discovery of the satellites to Mars in 1877." To all of us it is most interesting that during this year, when we are all sympathizing with the University of Padua in its celebration of the third centenary of its acquisition of Galileo as a Professor, we have first gained the knowledge of a fifth satellite in addition to the four discovered by Galileo.

EXTRACT FROM ADDRESS OF NOVEMBER 30, 1893.

Not the least important of the scientific events of the year is the publication, in the original German and in an English translation by Professor D. E. Jones, of a collection of Hertz's papers describing the researches by which he was led up to the experimental demonstration of magnetic waves. For this work the Rumford Medal of the Royal Society was delivered to Professor Hertz three years ago by my predecessor, Sir George Stokes. To fully appreciate the book now given to the world, we must carry our minds back to the early days of the Royal Society, when Newton's ideas regarding the forces which he saw to be implied in Kepler's laws of the motions of the Planets and of the Moon were frequent subjects of discussion at its regular meetings and at perhaps even more important non-official conferences among its fellows.

In 1684 the Senior Secretary of the Royal Society, Dr. Halley, went to Cambridge to consult Mr. Newton on the subject of the production of the elliptic motion of the Planets by a central force,[1] and on the 10th of December of that year he announced to the Royal Society that he " had seen Mr. Newton's book, *De Motu Corporum.*" Some time later, Halley was requested to "remind Mr. Newton of his promise to enter an account of his discoveries in the register of the Society," with the result that the great work *Philosophiæ Naturalis Principia Mathematica* was dedicated to the Royal Society, was actually presented in manuscript, and was communicated at an ordinary meeting of the Society on the 28th of April, 1686, by Dr. Vincent. In acknowledgment, it was ordered " that a letter of thanks be written to Mr.

---

[1] Whewell's *History of the Inductive Sciences,* vol. 2, p. 77.

Newton, and that the printing of his book be referred to the consideration of the Council ; and that in the meantime the book be put into the hands of Mr. Halley, to make a report thereof to the Council." On the 19th of May following, the Society resolved that " Mr. Newton's *Philosophiæ Naturalis Principia Mathematica* be printed forthwith in quarto, in a fair letter ; and that a letter be written to him to signify the Society's resolution, and to desire his opinion as to the volume, cuts, &c." An exceedingly interesting letter was accordingly written to Newton by Halley, dated London, May 22, 1686, which we find printed in full in Weld's *History of the Royal Society* (vol. 1, pp. 308—309). But the Council knew more than the Royal Society at large of its power to do what it wished to do. Biology was much to the front then, as now, and the publication of Willughby's book, *De Historia Piscium*, had exhausted the Society's finances to such an extent that the salaries even of its officers were in arrears. Accordingly, at the Council meeting of the 2nd of June, it was ordered that

M M 2

" Mr. Newton's book be printed, and that Mr. Halley undertake the business of looking after it, and printing it at his own charge, which he engaged to do."

It seems that at that time the office of Treasurer must have been in abeyance ; but with such a Senior Secretary as Dr. Halley there was no need for a Treasurer.

Halley, having accepted copies of Willughby's book, which had been offered to him in lieu of payment of arrears of salary[1] due to him, cheerfully undertook the printing of the *Principia*

---

[1] It is recorded in the Minutes of Council that the arrears of salary due to Hooke and Halley were resolved to be paid by copies of Willughby's work. Halley appears to have assented to this unusual proposition, but Hooke wisely " desired six months' time to consider of the acceptance of such payment."

The publication of the *Historia Piscium*, in an edition ot 500 copies, cost the Society £400. It is worthy of remark, as illustrative of the small sale which scientific books met with in England at this period, that a considerable time after the publication of Willughby's work, Halley was ordered by the Council to endeavour to effect a sale of several copies with a bookseller at Amsterdam, as appears in a letter from Halley requesting Boyle, then at Rotterdam, to do all in his power to give publicity to the book. When the Society resolved on Halley's undertaking to measure a degree of the Earth, it was voted that "he be given £50, or fifty *Books of Fishes*" (Weld's *History of the Royal Society*, vol. i., p. 310).

at his own expense, and entered instantly on the
duty of editing it with admirable zeal and energy,
involving, as it did, expostulations, arguments, and
entreaties to Newton not to cut out large parts of
the work which he wished to suppress[1] as being
too slight and popular, and as being possibly liable
to provoke questions of priority. It was well said
by Rigaud, in his "Essay on the first publication
of the *Principia*," that "under the circumstances,
it is hardly possible to form a sufficient estimate of
the immense obligation which the world owes in
this respect to Halley, without whose great zeal,
able management, unwearied perseverance, scientific attainments, and disinterested generosity,
the *Principia* might never have been published."[2]

---

[1] "The third [book] I now design to suppress. Philosophy is
such an impertinently litigious lady that a man had as good be
engaged in lawsuits as have to do with her. I found it so
formerly, and now I am no sooner come near her again but she
gives me warning. The first two books without the third will not
so well bear the title of *Philosophiæ Naturalis Principia Mathematica*, and therefore I have altered it to this, *De Motu Corporum
Libri duo;* but, upon second thoughts, I retain the former title.
'Twill help the sale of the book, which I ought not to diminish
now 'tis yours" (Weld's *History of the Royal Society*, vol. i. p. 311).

[2] *Ibid.*, p. 310.

Those who know how much worse than "law's delays" are the troubles, cares, and labour involved in bringing through the press a book on any scientific subject at the present day will admire Halley's success in getting the *Principia* published within about a year after the task was committed to him by the Royal Society, two hundred years ago.

When Newton's theory of universal gravitation was thus made known to the world Descartes's *Vortices*, an invention supposed to be a considerable improvement on the older invention of crystal cycles and epi-cycles from which it was evolved, was generally accepted, and seems to have been regarded as quite satisfactory by nearly all the philosophers of the day.

The idea that the Sun pulls Jupiter, and Jupiter pulls back against the Sun with equal force, and that the Sun, Earth, Moon, and Planets all act on one another with mutual attractions, seemed to violate the supposed philosophic principle that matter cannot act where it is not. Descartes's doctrine died hard among the mathematicians and philosophers of Continental Europe; and for the first quarter of last century belief in universal gravitation was an insularity of our countrymen.

Voltaire, during a visit which he made to England in 1727, wrote: "A Frenchman who arrives in London finds a great alteration in philosophy, as in other things. He left the world full; he finds it empty. At Paris you see the universe composed of vortices of subtle matter; at London we see nothing of the kind. With you it is the pressure of the Moon which causes the tides of the sea; in England it is the sea which gravitates towards the Moon. . . . You will observe also that the Sun, which in France has nothing to do with the business, here comes in for a quarter of it. Among you Cartesians all is done by impulsion: with the Newtonians it is done by an attraction of which we know the cause no better."[1] Indeed, the Newtonian opinions had scarcely any disciples in France till Voltaire asserted their claims on his return from England

[1] Whewell's *History of the Iuductive Sciences*, vol. 2, pp. 202—203.

in 1728. Till then, as he himself says, there were not twenty Newtonians out of England.[1]

In the second quarter of the century sentiment and opinion in France, Germany, Switzerland, and Italy experienced a great change. The mathematical prize questions proposed by the French Academy naturally brought the two sets of opinions into conflict. A Cartesian memoir of John Bernoulli was the one which gained the prize in 1730. It not unfrequently happened that the Academy, as if desirous to show its impartiality, divided the prize between Cartesians and Newtonians. Thus, in 1734, the question being the cause of the inclination of the orbits of the planets, the prize was shared between John Bernoulli, whose memoir was founded on the system of vortices, and his son Daniel, who was a Newtonian. The last act of homage of this kind to the Cartesian system was performed in 1740, when the prize on the question of the tides was distributed between Daniel Bernoulli, Euler, Maclaurin, and Cavallieri; the last of whom had tried to amend

[1] Whewell's *History of the Inductive Sciences*, vol. 2, p. 201.

and patch up the Cartesian hypothesis on this subject.[1]

On the 4th of February, 1744, Daniel Bernoulli wrote as follows to Euler: " Uebrigens glaube ich, dass der Aether sowohl *gravis versus solem*, als die Luft versus terram sey, und kann Ihnen night bergen, dass ich über diese Puncte ein völliger Newtonianer bin, vnd verwundere ich mich, dass sie den Principiis Cartesianis so lang adhariren; es möchte wohl einige Passion vielleicht mit unterlaufen. Hat Gott können eine *animam*, deren Natur uns unbegreiflich ist, erschaffen, so hat er auch können eine attractionem universalem materiae imprimiren, wenn gleich solche attractio *supra captum* ist, da hingegen die Principia Cartesiana allzeit *contra captum* etwas involviren."

Here the writer, expressing wonder that Euler had so long adhered to the Cartesian principles, declares himself a thorough-going Newtonian, not merely in respect to gravitation *versus* vortices, but in believing that matter may have been created simply with the law of universal attraction without

[1] Whewell's *History of the Inductive Sciences*, vol. 2, pp. 198, 199.

the aid of any gravific medium or mechanism. But in this he was more Newtonian than Newton himself.

Indeed Newton was not a Newtonian, according to Daniel Bernoulli's idea of Newtonianism, for in his letter to Bentley of date 25th February, 1792,[1] he wrote: "That gravity should be innate, inherent, and essential to matter, so that one body may act upon another at a distance through a vacuum without the mediation of anything else, by and through which their action and force may be conveyed from one to another, is to me so great an absurdity that I believe no man who has in philosophical matters a competent faculty of thinking can ever fall into it." Thus Newton, in giving out his great law, did not abandon the idea that matter cannot act where it is not. In respect, however, merely of philosophic thought, we must feel that Daniel Bernoulli was right; we can conceive the Sun attracting Jupiter, and Jupiter attracting the Sun, without any intermediate medium, if they are ordered to do so. But the question remains—Are

---

[1] *The Correspondence of Richard Bentley, D.D.*, vol. I, p. 70.

they so ordered? Nevertheless, I believe all, or nearly all, his scientific contemporaries agreed with Daniel Bernoulli in answering this question affirmatively. Very soon after the middle of the eighteenth century Father Boscovich[1] gave his brilliant doctrine (if infinitely improbable theory) that elastic rigidity of solids, the elasticity of compressible liquids and gases, the attractions of chemical affinity and cohesion, the forces of electricity and magnetism—in short, all the properties of matter except heat, which he attributed to a sulphureous fermenting essence—are to be explained by mutual attractions and repulsions, varying solely with distances, between mathematical points endowed also, each of them, with inertia. Before the end of the eighteenth century the idea of action-at-a-distance through absolute vacuum had become so firmly established, and Boscovich's theory so unqualifiedly accepted as a reality, that the idea of gravitational force or

[1] *Theoria Philosophiæ Naturalis Redacta ad unicam legem virium in natura existentium auctore P. Rogerio Josepho Boscovich, Societatis Jesu,* 1st edition, Vienna, 1758; 2nd edition, amended and extended by the author, Venice, 1763.

electric force or magnetic force being propagated through and by a medium seemed as wild to the naturalists and mathematicians of 100 years ago as action-at-a-distance had seemed to Newton and his contemporaries 100 years earlier. But a retrogression from the eighteenth century school of science set in early in the nineteenth century.

Faraday, with his curved lines of electric force, and his dielectric efficiency of air and of liquid and solid insulators, resuscitated the idea of a medium through which, and not only through which but *by* which, forces of attraction or repulsion, seemingly acting at a distance, are transmitted. The long struggle of the first half of the eighteenth century was not merely on the question of a medium to serve for gravific mechanism, but on the correctness of the Newtonian law of gravitation as a matter of fact however explained. The corresponding controversy in the nineteenth century was very short, and it soon became obvious that Faraday's idea of the transmission of electric force by a medium not only did not violate Coulomb's law of relation between force and distance, but that, if real, it must give a thorough explanation of that law.[1] Nevertheless, after Faraday's discovery [2] of the different specific inductive capacities of different insulators, twenty years passed before it was generally accepted in Continental Europe. But before his death, in 1867, he had succeeded in inspiring the rising generation of the scientific world with something approaching to faith that electric force is transmitted by a medium called ether, of which, as had been believed by the whole scientific world for forty years, light and radiant heat are transverse vibrations. Faraday himself did not rest with this theory for electricity alone. The very last time I saw him at work in the Royal Institution was in an underground cellar, which he had chosen for freedom from disturbance; and he was arranging experiments to test the time of propagation of magnetic force from an electro-magnet through a distance of many yards of air to a fine steel needle polished to reflect

---

[1] *Electrostatics and Magnetism,* Sir W. Thomson, Arts. I. (1842) and II. (1845), particularly § 25 of Art. II.

[2] 1837, *Experimental Researches,* 1161—1306.

light ; but no result came from those experiments. About the same time or soon after, certainly not long before the end of his working time, he was engaged (I believe at the shot tower near Waterloo Bridge on the Surrey side) in efforts to discover relations between gravity and magnetism, which also led to no result.

Absolutely nothing has hitherto been done for gravity either by experiment or observation towards deciding between Newton and Bernoulli, as to the question of its propagation through a medium, and up to the present time we have no light, even so much as to point a way for investigation, in that direction. But for electricity and magnetism, Faraday's anticipations and Clerk-Maxwell's splendidly developed theory have been established on the sure basis of experiment by Hertz's work, of which his own most interesting account is this year presented to the world in the German and English volumes to which I have referred. It is interesting to know, as Hertz explains in his introduction, and it is very important in respect to the experimental demonstration of magnetic waves to which he was led, that he began his electric researches in a problem happily put before him thirteen years ago by Professor von Helmholtz, of which the object was to find by experiment some relation between electromagnetic forces and dielectric polarisation of insulators, without, in the first place, any idea of discovering a progressive propagation of those forces through space.

It was by sheer perseverance in philosophical experimenting that Hertz was led to discover a finite velocity of propagation of electromagnetic action, and then to pass on to electromagnetic waves in air and their reflection, and to be able to say, as he says in a short reviewing sentence at the end of his eighth paper : " Certainly it is a fasci- " nating idea that the processes in air which we "have been investigating, represent to us on a " million-fold larger scale the same processes " which go on in the neighbourhood of a Fresnel " mirror, or between the glass plates used for " exhibiting Newton's rings."

Professor Oliver Lodge has done well, in connexion with Hertz's work, to call attention[1] to old experiments, and ideas taken from them, by Joseph Henry, which came more nearly to an experimental demonstration of electromagnetic waves than anything that had been done previously. Indeed Henry, after describing experiments showing powerful enough induction due to a single spark from the prime conductor of an electric machine to magnetise steel needles at a distance of thirty feet in a cellar beneath with two floors and ceilings intervening, says that he is "disposed to adopt the hypothesis of an electrical plenum," and concludes with a short reviewing sentence: "It may be further inferred that the diffusion of "motion in this case is almost comparable with "that of a spark from a flint and steel in the case "of light."

Professor Oliver Lodge himself did admirable work in his investigations with reference to lightning rods,[2] coming very near to experimental demonstrations of electromagnetic waves; and he

[1] *Modern Views of Electricity,* pp. 369—372.

[2] *Lightning Conductors and Lightning Guards,* Oliver J. Lodge, D.Sc., F.R.S. Whittaker and Co.

drew important lessons regarding "electrical surgings" in an insulated bar of metal "induced "by Maxwell's and Heaviside's electromagnetic "waves," and many other corresponding phenomena manifested both in ingenious and excellent experiments devised by himself and in natural effects of lightning.

Of electrical surgings or waves in a short insulated wire, and of interference between ordinary and reflected waves, and positive electricity appearing where negative might have been expected, we hear first, it seems, in Herr von Bezold's "Researches on the Electric Discharge" (1870), which Hertz gives as the third paper of his collection, with interesting and ample recognition of its importance in relation to his own work.

In connexion with the practical development of magnetic waves, you will, I am sure, be pleased if I call your attention to two papers by Professor G. F Fitzgerald, which I heard myself at the meeting of the British Association at Southport, in 1883. One of them is entitled "On a Method of Producing Electromagnetic Disturbances of comparatively Short Wave-lengths." The paper itself is not long, and I shall read it to you in full, from the *Report of the British Association*, 1883: " This is by utilising the alternating currents produced when an accumulator is discharged through a small resistance. It is possible to produce waves of as little as 2 metres wave-length, or even less." This was a brilliant and useful suggestion. Hertz, not knowing of it, used the method; and, making as little as possible of the "accumulator," got waves of as little as 24 cm. wave-length in many of his fundamental experiments. The title alone of Fitzgerald's other paper, " On the Energy Lost by Radiation from Alternating Currents," is in itself a valuable lesson in the electromagnetic theory of light, or the undulatory theory of magnetic disturbance. It is interesting to compare it with the title of Hertz's eleventh paper, " Electric Radiation " ; but I cannot refer to this paper without expressing the admiration and delight with which I see the words "rectilinear propagation," " polarisation," " reflection," " refraction," appearing in it as subtitles.

During the fifty-six years which have passed since Faraday first offended physical mathematicians with his curved lines of force, many workers and many thinkers have helped to build up the nineteenth century school of *plenum;* one ether for light, heat, electricity, magnetism ; and the German and English volumes containing Hertz's electrical papers, given to the world in the last decade of the century, will be a permanent monument of the splendid consummation now realised.

But, splendid as this consummation is, we must not fold our hands and think or say there are no more worlds to conquer for electrical science. We do know something now of magnetic waves. We know that they exist in nature and that they are in perfect accord with Maxwell's beautiful theory. But this theory teaches us nothing of the actual motions of matter constituting a magnetic wave. Some definite motion of matter perpendicular to the lines of alternating magnetic force in the waves and to the direction of propagation of the action through space, there must be ; and it seems almost satisfactory as a

N N 2

hypothesis to suppose that it is chiefly a motion of ether with a comparatively small but not inconsiderable loading by fringes of ponderable molecules carried with it. This makes Maxwell's "electric displacement" simply a to-and-fro motion of ether across the line of propagation, that is to say, precisely the vibrations in the undulatory theory of light according to Fresnel. But we have as yet absolutely no guidance towards any understanding or imagining of the relation between this simple and definite alternating motion, or any other motion or displacement of the ether, and the earliest known phenomena of electricity and magnetism—the electrification of matter, and the attractions and repulsions of electrified bodies; the permanent magnetism of lodestone and steel, and the attractions and repulsions due to it : and certainly we are quite as far from the clue to explaining, by ether or otherwise, the enormously greater forces of attraction and repulsion now so well known after the modern discovery of electromagnetism.

Fifty years ago it became strongly impressed on

my mind that the difference of quality between vitreous and resinous electricity, conventionally called positive and negative, essentially ignored as it is in the mathematical theories of electricity and magnetism with which I was then much occupied (and in the whole science of magnetic waves as we have it now), must be studied if we are to learn anything of the nature of electricity and its place among the properties of matter. This distinction, essential and fundamental as it is in frictional electricity, electro-chemistry, thermo-electricity, pyro-electricity of crystals, and piezo-electricity of crystals, had been long observed in the old known beautiful appearances of electric glow and brushes and sparks from points and corners on the con-ductors of ordinary electric machines and in exhausted receivers of air pumps with electricity passed through them. It was also known, probably as many as fifty years ago, in the vast difference of behaviour of the positive and negative electrodes of the electric arc lamp. Faraday gave great attention to it in experiments and observations

[1] *Experimenta. Researches*, Series 12 and 13, Jan. and Feb., 1838.

regarding electric sparks, glows, and brushes, and particularly in his "dark discharge" and "dark space" in the neighbourhood of the negative electrode in partial vacuum. In [1523] of his 12th series, he says, "The results connected with the different conditions of positive and negative discharge will have a far greater influence on the philosophy of electrical science than we at present imagine." His "dark discharge" ([1544—1554]) through space around or in front of the negative electrode was a first instalment of modern knowledge in that splendid field of experimental research which, fifteen years later, and up to the present time, has been so fruitfully cultivated by many of the ablest scientific experimenters of all countries.

The Royal Society's *Transactions* and *Proceedings* of the last forty years contain, in the communications of Gassiot,[1] Plücker,[2] Andrews and Tait,[3] Robinson,[4] Cromwell Varley,[5] De la Rue

[1] *Roy. Soc. Proc.*, vol. 10, 1860, pp. 36, 269, 274, 432.
[2] *Ibid.* p. 256.
[3] *Ibid*, p. 274; *Phil. Trans.* 1860, p. 118.
[4] *Roy. Soc. Proc.*, vol. 12, 1862, p. 202.
[5] *Ibid.* vol. 19, 1871, p. 236.

and Müller,[1] Spottiswoode,[2] Moulton,[3] Grove,[4] Crookes,[5] Schuster,[6] J. J. Thomson[7], and Fleming,[8] almost a complete history of the new province of electrical science which has grown up, largely in virtue of the great modern improvements in practical methods for exhausting air from glass vessels, culminating in Sprengel's mercury-shower pump, by which we now have "vacuum tubes" and bulbs containing less than $\frac{1}{190,000}$ of the air which would be left in them by all that

---

[1] *Roy. Soc. Proc.*, vol. 23, 1875, p. 356; vol. 26, 1877, p. 519; vol. 27, 1878, p. 374; vol. 29, 1879, p. 281; vol. 35, 1883, p. 292; vol. 36, 1884, pp. 151, 206; *Phil. Trans.*, 1878, pp. 55, 155; 1880, p. 65; 1883, 477.

[2] *Roy. Soc. Proc.*, vol. 23, 1875, pp. 356, 455; vol. 25, 1875, pp. 73, 547; vol. 26, 1877, pp. 90, 323; vol. 27, 1878, p. 60; vol. 29, 1879, p. 21; vol. 30, 1880, p. 302; vol. 32, 1881, pp. 385, 388; vol. 33, 1882, p. 423; *Phil. Trans.*, 1878, pp. 163, 210; 1879, 165; 1880, p. 561.

[3] *Roy. Soc. Proc.*, vol. 29, 1879, p. 21; vol. 30, 1880, p. 302; vol. 32, 1881, pp. 385, 388; vol. 33, 1882, p. 453; *Phil. Trans.*, 1879, p. 165; 1880, p. 561.

[4] *Roy. Soc. Proc.* vol. 28, 1878, p. 181.

[5] *Ibid.* 1879, pp. 347, 477; *Phil. Trans.*, 1879, p. 641; 1880, p. 1 5; 1881, 387.

[6] *Roy. Soc. Proc.*, vol. 37, 1884, pp. 78, 317; vol. 42, 1887, p. 371; vol. 47, 1890; pp. 300, 506.

[7] *Ibid.* vol. 42, 1887, p. 343; vol. 49, 1891, p. 84.

[8] *Ibid.* vol. 47, 1890, p. 118.

could be done in the way of exhausting (supposed to be down to 1 mm of mercury) by the best air-pump of fifty years ago. A large part of the fresh discoveries in this province has been made by the authors of these communications; and their references to the discoveries of other workers very nearly complete the history of all that has been done in the way of investigating the transmission of electricity through highly rarefied air and gases since the time of Faraday.

Varley's short paper of 1871, which, strange to say, has lain almost or quite unperceived in the *Proceedings* during the twenty-two years since its publication, contains an important first instalment of discovery in a new field—the molecular torrent from the "negative pole," the control of its course by a magnet, its pressure against either end of a pivoted vane of mica according as it is directed by a magnet to one end or the other, and the shadow produced by its interception by a mica screen. Quite independently of Varley, and not knowing what he had done, Crookes was led to the same primary discovery, not by accident, and not merely by experimental skill and acuteness of observation. He was led to it by carefully designed investigation, starting with an examination of the cause of irregularities which had troubled[1] him in his weighing of thallium ; and, going on to trials for improving Cavendish's gravitational measurement, in the course of which he discovered that the seeming attraction by heat is only found in air of greater than $\frac{1}{1000}$ of ordinary density ;[2] and that there is repulsion increasing to a maximum when the density is decreased from $\frac{1}{1000}$ to $\frac{36}{1,000,000}$ and thence diminishing towards zero as the rarefaction is farther extended to density $\frac{1}{20,000,000}$. From this discovery Crookes came to his radiometer, first without and then with electrification ; and, powerfully aided by Sir George Stokes,[3] he brought all his work more and more into touch with the kinetic theory of gases ; so

[1] Tribulation, not undisturbed progress, gives life and soul, and leads to success when success can be reached, in the struggle for natural knowledge.

[2] Crookes, "On the Viscosity of Gases at High Exhaustions, § 655, *Phil. Trans.*, Feb. 1881, p. 403.

Ibid., vol. 172 (1881), pp. 387, 435.

much so that when he discovered the molecular torrent he immediately gave it its true explanation —molecules of residual air, or gas, or vapour projected at great velocities[1] by electric repulsion from the negative electrode. This explanation has been repeatedly and strenuously attacked by many other able investigators, but Crookes has defended[2] it, and thoroughly established it by what I believe is irrefragable evidence of experiment. Skilful investigation perseveringly continued brought out more and more of wonderful and valuable results : the non-importance of the position of the positive electrode ; the projection of the torrent *perpendicularly* from the surface of the negative electrode ; its convergence to a focus and divergence thenceforward when the surface is slightly convex ; the slight but perceptible repulsion between two parallel torrents due, according to Crookes, to negative electrifications of their constituent molecules ; the change of direction of

---

[1] Probably, I believe, not greater in any case than two or three kilometres per second.

[2] Address to the Institution of Electrical Engineers, 1891.

the molecular torrent by a neighbouring magnet ;
the tremendous heating effect of the torrent from
a concave electrode when glass, metal, or any
ponderable substance is placed in the focus ; the
phosphorescence produced on a plate coated with
sensitive paint by a molecular torrent skirting
along it ; the brilliant colours—turquoise-blue,
emerald, orange, ruby-red—with which gray
colourless objects and clear colourless crystals glow
on their struck faces when lying separately or piled
up in a heap in the course of a molecular torrent ;
" electrical evaporation " of negatively electrified
liquids and solids ;[1] the seemingly red hot glow
but with no heat conducted inwards from the
surface, of cool solid silver kept negatively electrified in a vacuum of $\frac{1}{1,000,000}$ of an atmosphere,
and thereby caused to rapidly evaporate. This
last-mentioned result is almost more surprising
than the phosphorescent glow excited by molecular
impacts in bodies not rendered perceptibly phosphorescent by light. Both phenomena will surely
be found very telling in respect to the molecular

[1] *Roy. Soc. Proc.*, June 11, 1891.

constitution of matter and the origination of thermal radiation, whether visible as light or not. In the whole train of Crookes' investigations on the radiometer, the viscosity of gases at high exhaustions, and the electric phenomena of high vacuums, ether seems to have nothing to do except the humble function of showing to our eyes something of what the atoms and molecules are doing. The same confession of ignorance must be made with reference to the subject dealt with in the important researches of Schuster and J. J. Thomson on the passage of electricity through gases. Even in Thomson's beautiful experiments showing currents produced by circuital electromagnetic induction in complete poleless circuits, the presence of molecules of residual gas or vapour seems to be *the essential*. It seems certainly true that without the molecules there could be no current, and that without the molecules electricity has no meaning. But in obedience to logic I must withdraw one expression I have used. We must not imagine that "presence of molecules is *the* essential." It is certainly *an* essential. Ether also is certainly *an* essential, and certainly has more to do than merely to telegraph to our eyes to tell us of what the molecules and atoms are about. If a first step towards understanding the relations between ether and ponderable matter is to be made, it seems to me that the most hopeful foundation for it is knowledge derived from experiment on electricity in high vacuum ; and if, as I believe is true, there is good reason for hoping to see this step made, we owe a debt of gratitude to the able and persevering workers of the last forty years who have given us the knowledge we have: and we may hope for more and more from some of themselves and from others encouraged by the fruitfulness of their labours to persevere in the work.

ADDRESS

[*Delivered on the occasion of the unveiling of Joule's statue in Manchester Town Hall, December 7th,* 1893.]

I THANK the Committee for the great honour it has done me in asking me to be present upon an occasion so full of interest to the city of Manchester, and certainly most interesting to myself personally. The proceedings which have just taken place have given Manchester the possession of a work of art which will remain an ornament and an honour to the city. I am afraid if I were to say even a small part of what I feel upon this occasion I should tax the patience of my audience to an intolerable degree   At the same time I believe you would all wish to hear something of Joule's work.

Joule's work began in Manchester, was carried on in Manchester, and finished in Manchester.   It began very early, when he was only nineteen years of age.   He was not altogether a self-taught man in science.   After a good ordinary school education, he had the inestimable benefit of the personal teaching of Dalton in chemistry. He and his elder brother Benjamin were favourite pupils of Dalton. They went to his house in the rooms of the Literary and Philosophical Society of Manchester for regular daily lessons and were a little disappointed at first when they found that Dalton, instead of introducing them straight away to the grandeur of the atomic theory of chemistry, kept them to the grindstone, forced them to do their additions correctly, and held up to them as something essentially necessary for them to learn, the practice of trigonometry and the logarithmic tables. James Joule and his brother got great good from that early severe, almost hard, training by Dalton. They were both full of original brightness and acuteness in their observations. They went through the country even before they came to be pupils of Dalton, making memoranda of what they saw and heard, an aurora borealis or a wonderful thunderstorm, or sounds of artillery or lightning, they could not tell which. Some of their journals they afterwards showed to

Dalton, who thought so well of their descriptions that in one instance he was able to say to them, "Those sounds you heard were not human artillery but they were the thunder of an outburst of lightning at sea forty miles south of Holyhead." The two brothers continued pupils of Dalton until the failure of his health ; but for a year after that, and no doubt to the very end, they continued to receive ideas from that great man. It must not be thought that Dalton only taught them arithmetic and trigonometry. I rather emphasize that point with an eye perhaps to the young men who aspire to follow in Joule's footsteps, and upon whom I wish to impress the conviction that it was hard work early begun and persevered in and conscientiously carried out ; that is the foundation of all great works, whether in literature, philosophy, or science, or in doing good to the world in any possible way. In electricity and electro-magnetism Joule, I think I may say, was wholly self-taught. All he knew he learned from his own reading—from reading in text books and in Sturgeon's *Annals of Electricity*, and also from conferences

with Sturgeon himself. The Literary and Philosophical Society of Manchester has the distinguished
honour of having been the cradle of Joule's
scientific childhood when it was Dalton's home,
and of being afterwards Joule's life-long scientific
harbour. From those early days he kept constantly
in touch with that Society. Many of his most
important papers were first given to the world
there, and during the last years of his life he was
an exceedingly regular, it might almost be said a
constant, attendant at the meetings of the Society.

An interesting and sympathetic memoir of
Joule, with much important scientific information
and judgment regarding his work, by Professor
Osborne Reynolds, constitutes the sixth volume of
the fourth series of its "Memoirs and Proceedings."

The citizens of Manchester do not require to be
told what great things their Literary and Philosophical Society in its rather more than a
century's existence has done for them and for the
world. Your being here in such numbers on the
present occasion shows how much you appreciate
the results of that very effective scientific institu-

O O

tion. Now I ought to say something of the electrical, mechanical, and chemical character of Joule's work, although to examine it properly would require the space not of one short address but of a whole course of lectures illustrated by experiment. A great surprise that came out very early in Joule's work was burning without heat—an absolutely novel idea which Joule developed most wonderfully and most magnificently by his experiments on the generation of heat in the voltaic battery. Joule was the first to develop the idea, and it came to him not as a bright flash of genius, but as the demonstrated result of years of hard, measuring, calculating work. This burning without heat was a fundamental idea that pervaded all Joule's work. A few years later he expanded it in an admirable way. About 1844, in a joint paper by himself and Scoresby, "On the Mechanical Powers of Electromagnetism, Steam and Horses," he brought out the startling but truly philosophical idea that when a man or any other animal walked uphill only a part of the heat of combustion of his food was developed, and that it was only when the body was quiescent or walking about on a level or going downhill that the chemical attraction between the food and the oxygen dissolved in the blood developed its whole energy in animal heat. He showed, further, that a horse or a man employed in doing mechanical work against resistance was more economical of fuel than was any steam engine hitherto realised. This was a very far-reaching idea, and seemed to hold out prospects of greatly advancing the efficiency of the steam engine. That promise has not been lost. It is due to Joule more than to any other individual that the great improvement of surface condensation was now universal; although it was very rarely practised before 1860 or 1862. Between 1855 and 1862, Joule and I had a small steam engine fitted up in the stable of his father's house, Oakfield, for use in our joint investigations on the thermic effects on fluids in motion. To that little steam engine Joule applied a surface condenser on an entirely new principle and plan, which gave us such good results that starting from it he undertook a special investigation on the surface-condensation of steam with the assistance of a grant from the Royal Society for the purpose. The results of this very elaborate investigation, communicated to the Royal Society on December 13, 1860, and published in the *Philosophical Transactions*, have proved to be of enormous practical importance. They led directly and speedily to the present practical method of surface-condensation which is one of the most valuable improvements of the steam engine, especially for marine use, since the time of Watt.

But I have not yet touched upon Joule's great fundamental discovery, the discovery which is first in every one's mouth when asked what was Joule's work?—The Mechanical Equivalent of Heat. You must understand that it was not merely by a chance piece of experiment or of guessing that he stumbled on a result which was afterwards found to be of great value. It was measurement, rigorous experiment and observation, and philosophic thought all round the field of physical science that made the discovery possible to him.

Very early, however, in his working time Joule
brought out the mechanical equivalent of heat,
and in a paper at the British Association at Cork
in 1843, published afterwards in the *Philosophical
Magazine*, he gave the number " 770." Six years
later a second determination gave him a result
about ⅛ per cent. larger, and twenty-nine years
later he completed a third determination. The
result of this final investigation of Joule's is 772·43
Manchester foot lbs. for the quantity of heat required to warm from 60° to 61° Fahr., one pound
of water weighed in vacuum: which is about $\frac{1}{20}$ per
cent. greater than the result of 1849 expressed in
the same term.

In the year 1824 a great theory was originated
by a very young man, who died only a few years
later—Sadi Carnot, son of the Republican War
Minister and uncle of the present President of the
French Republic. It was he who made " Carnot's
theory" a household word throughout the world
of science ; and great as is the French President,
much as he has done and is doing for his country
and the world, in after times his uncle Sadi will be

always remembered as one of the most illustrious members of that great family. Carnot's theory gave an important fundamental principle regarding the development of motive power from heat. Joule's work, on the other hand, so far as the mechanical equivalent was concerned, was the generation of heat by mechanical work. It was quite the middle of the century before Carnot's theory began to attract attention ; but Joule was early made acquainted with it, and after fighting a little against it, as differing from his own theory, he of all others took it up in the most hearty manner. I can never forget the British Association at Oxford in the year 1847, when in one of the sections I heard a paper read by a very unassuming young man who betrayed no consciousness in his manner that he had a great idea to unfold. I was tremendously struck with the paper. I at first thought it could not be true because it was different from Carnot's theory, and immediately after the reading of the paper I had a few words of conversation with the author James Joule, which was the beginning of our forty years' acquaintance and friendship. On the evening of the same day that very valuable Institution of the British Association, its conversazione, gave us opportunity for a good hour's talk and discussion over all that either of us knew of thermodynamics. I gained ideas which had never entered my mind before, and I thought I too suggested something worthy of Joule's consideration when I told him of Carnot's theory. Then and there in the Radcliffe Library, Oxford, we parted, both of us, I am sure, feeling that we had much more to say to one another and much matter for reflection in what we had talked over that evening. But what was my surprise a fortnight later when, walking down the valley of Chamounix, I saw in the distance a young man walking up the road towards me and carrying in his hand something which looked like a stick, but which he was using neither as an Alpenstock nor as a walking stick. It was Joule with a long thermometer in his hand, which he would not trust by itself in the *char-à-banc* coming slowly up the hill behind him lest it should get broken. But there comfortably and safely seated on the *char-à-banc* was his bride—the sympathetic companion and sharer in his work of after years. He had not told me in Section A or in the Radcliffe Library that he was going to be married in three days, but now in the valley of Chamounix, he introduced me to his young wife. We appointed to meet again a fortnight later at Martigny to make experiments on the heat of a waterfall (Sallanches) with that thermometer: and afterwards we met again and again and again, and from that time indeed remained close friends till the end of Joule's life. I had the great pleasure and satisfaction for many years, beginning just forty years ago, of making experiments along with Joule which led to some important results in respect to the theory of thermodynamics. This is one of the most valuable recollections of my life, and is indeed as valuable a recollection as I can conceive in the possession of any man interested in science. Joule's initial work was the very foundation of our knowledge of the steam engine and steam power. Taken along with Carnot's theory it has given the scientific foundation on which all the great improvements since the

year 1850 have been worked out, not in a haphazard
way but on a careful philosophical basis. James
Watt had anticipated to some degree in his compound engine and his expansive system the benefits
now realised, but he was before his time in that
respect and he had not the complete foundation
which Joule's mechanical equivalent and Carnot's
theory have since given for the improvement of the
steam engine.

May I be allowed to congratulate the city of
Manchester on its proceedings to-day? When
the cover was lifted from the statue of Joule
I felt deeply touched at the sight of the face
of my old friend. To my mind it is a most
admirable likeness, and the ideality of the ac-
cessory of the little brass model held in the hand,
the eidolon of what was in the mind of the
powerful thinking face shown in marble, seems to
me most interesting and most striking—I think I
may say poetical. This little model is not Joule's
first apparatus nor his second: it is his third and
greatest apparatus for the determination of the
mechanical equivalent of heat—that by which he

corrected the British Association's standard ohm, which he found to be 1·7 per cent. wrong. Regarding the ohm a diplomatic correspondence is now going on through our Foreign Office with other Governments for the purpose of arranging the precise terms of the definition of the ohm, of which a correct standard was really first worked out by Joule. May I be allowed to congratulate the sculptor, Mr. Gilbert, on the great beauty, originality, and success of his work. Manchester now possesses two statues, Dalton on the left and Joule on the right of the entrance to its Municipal Buildings ; the man who laid the foundation of the atomic theory in chemistry and the man who discovered the mechanical equivalent of heat. If the prosperity of Manchester does not depend on chemistry and on the steam engine and thermo-dynamics I do not know upon what it does depend, unless it be the energy and industry and honourable character of its inhabitants ; but you must ever remember that the material prosperity of this great city has owed more to philosophic thought than to any material appliance whatever.

ISOPERIMETRICAL PROBLEMS.

[Being a Friday evening Lecture delivered to the Royal Institution, May 12th, 1893.]

Dido, B.C. 800 or 900.
Horatius Cocles, B.C. 508.
Pappus, Book V., A.D. 390.
John Bernoulli, A.D. 1700.
Euler, A.D. 1744.
Maupertuis (Least Action), b. 1698, d. 1759.
Lagrange (Calculus of Variations), 1759.
Hamilton (Actional Equations of Dynamics), 1834.
Liouville, 1840 to 1860.

THE first isoperimetrical problem known in history was practically solved by Dido, a clever Phœnician princess, who left her Tyrian home and emigrated to North Africa, with all her property and a large retinue, because her brother Pygmalion murdered her rich uncle and husband Acerbas, and plotted to defraud her of the money which he left. On landing in a bay about the middle of the north coast of Africa she obtained a grant from Hiarbas, the native chief of the district, of as much land as she could enclose with an ox-hide. She cut the ox-hide into an exceedingly long strip, and succeeded in enclosing between it and the sea a very valuable territory[1] on which she built Carthage.

The next isoperimetrical problem on record was three or four hundred years later, when Horatius Cocles, after saving his country by defending the bridge until it was destroyed by the Romans behind him, saved his own life and got back into Rome by swimming the Tiber under the broken bridge, and was rewarded by his grateful countrymen with a grant of as much land as he could plough round in a day.

In Dido's problem the greatest value of land was to be enclosed by a line of given length.

---

[1] Called Byrsa, from βύρσα, the hide of a bull. [Smith's *Dictionary of Greek and Roman Biography and Mythology,* article "Dido."]

If the land is all of equal value the general
solution of the problem shows that her line of
ox-hide should be laid down in a circle. It
shows also that if the sea is to be part of the
boundary, starting, let us say, southward from any

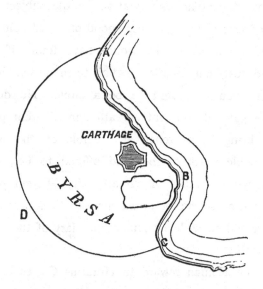

given point, A, of the coast, the inland bounding
line must at its far end cut the coast line
perpendicularly. Here, then, to complete our
solution, we have a very curious and interesting,
but not at all easy, geometrical question to

answer :—What must be the radius of a circular arc, A D C, of given length, and in what direction must it leave the point A, in order that it may cut a given curve, A B C, perpendicularly at some unknown point, C? I don't believe Dido could have passed an examination on the subject, but no doubt she gave a very good practical solution, and better than she would have found if she had just mathematics enough to make her fancy the boundary ought to be a circle. No doubt she gave it different curvature in different parts to bring in as much as possible of the more valuable parts of the land offered to her, even though difference of curvature in different parts would cause the total area enclosed to be less than it would be with a circular boundary of the same length.

The Roman reward to Horatius Cocles brings in quite a new idea, now well known in the general subject of isoperimetrics : the greater or less speed attainable according to the nature of the country through which the line travelled over passes. If it had been equally easy to

plough the furrow in all parts of the area
offered for enclosure, and if the value of the
land per acre was equal throughout, Cocles would
certainly have ploughed as nearly in a circle as
he could, and would only have deviated from
a single circular path if he found that he had
misjudged its proper curvature. Thus, he might
find that he had begun on too large a circle
and, in order to get back to the starting point
and complete the enclosure before nightfall, he
must deviate from it on the concave side; or he
would deviate from it on the other side if he
found that he had begun on too small a circle
and that he had still time to spare for a wider
sweep. But, in reality, he must also have considered the character of the ground he had to
plough through, which cannot but have been
very unequal in different parts, and he would
naturally vary the curvature of his path to
avoid places where his ploughing must be very
slow, and to choose those where it would be most
rapid.

He must also have had, as Dido had, to consider the different value of the land in different parts, and thus he had a very complex problem to practically solve. He had to be guided both by the value of the land to be enclosed and the speed at which he could plough according to the path chosen; and he had a very brain-trying task to judge what line he must follow to get the largest value of land enclosed before night.

These two very ancient stories, whether severe critics will call them mythical or allow them to be historic, are nevertheless full of scientific interest. Each of them expresses a perfectly definite case of the great isoperimetrical problem to which the whole of dynamics is reduced by the modern mathematical methods of Euler, Lagrange, Hamilton, and Liouville (Liouville's Journal, 1840–1850). In Dido's and Horatius Cocles' problems, we find perfect illustrations of all the fundamental principles and details of the generalised treatment of dynamics which we have learned from these great mathematicians of the eighteenth and nineteenth centuries.

Nine hundred years after the time of Horatius

Cocles we find, in the fifth Book of the collected
Mathematical and Physical Papers of Pappus
of Alexandria, still another idea belonging to
isoperimetrics—the economy of valuable material
used for building a wall; which, however, is
virtually the same as the time per yard of furrow
in Cocles' ploughing. In this new case the
economist is not a clever princess, nor a patriot
soldier; but a humble bee who is praised in the
introduction to the book not only for his
admirable obedience to the Authorities of his
Republic, for the neat and tidy manner in
which he collects honey, and for his prudent
thoughtfulness in arranging for its storage and
preservation for future use, but also for his
knowledge of the geometrical truth that a
"hexagon can enclose more honey than a square
or a triangle with equal quantities of building
material in the walls," and for his choosing
on this account the hexagonal form for his
cells. Pappus, concluding his introduction with
the remark that bees only know as much of
geometry as is practically useful to them,

proceeds to apply what he calls his own superior human intelligence to investigation of useless knowledge, and gives results in his Book V., which consists of fifty-five theorems and fifty-seven propositions on the areas of various plane figures having equal circumferences. In this Book, written originally in Greek, we find (Theorem IX. Proposition X.) the expression "isoperimetrical figures," which is, so far as I know, the first use of the adjective "isoperimetrical" in geometry; and we may, I believe, justly regard Pappus as the originator, for mathematics, of *isoperimetrical problems*, the designation technically given in the nineteenth century[1] to that large province of mathematical and engineering science in which different figures having equal circumferences, or different paths between two given points, or between some two points on two given curves, or on one given curve, are compared in connection with definite questions of greatest efficiency and smallest cost.

[1] Example, Woodhouse's *Isoperimetrical Problems*, Cambridge, 1810.

In the modern engineering of railways an isoperimetrical problem of continual recurrence is the laying out of a line between two towns along which a railway may be made at the smallest prime cost. If this were to be done irrespectively of all other considerations, the requisite datum for its solution would be simply the cost per yard of making the railway in any part of the country between the two towns. Practically the solution would be found in the engineers' drawing office by laying down two or three trial lines to begin with, and calculating the cost of each, and choosing the one of which the cost is least. In practice various other considerations than very slight differences in the cost of construction will decide the ultimate choice of the exact line to be taken, but if the problem were put before a capable engineer to find very exactly the line of minimum total cost, with an absolutely definite statement of the cost per yard in every part of the country, he or his draughtsmen would know perfectly how to find the solution. Having found something near the true line by a few rough trials they would try small deviations from the rough approximation, and calculate differences of cost for different lines differing very little from one another. From their drawings and calculations they would judge by eye which way they must deviate from the best line already found, to find one still better. At last they would find two lines for which their calculation shows no difference of cost. Either of these might be chosen; or, according to judgment, a line midway between them, or somewhere between them, or even not between them but near to one of them, might be chosen, as the best approximation to the exact solution of the mathematical problem which they care to take the labour of trying for. But it is clear that if the price per yard of the line were accurately given (however determined or assumed) there would be an absolutely definite solution of the problem, and we can easily understand that the skill available in a good engineer's drawing-office would suffice to find the solution with any degree of accuracy that might be prescribed; the minuter the accuracy to be attained the greater the labour, of course. You must not imagine that I suggest, as a thing of practical engineering, the attainment of minute accuracy in the solution of a problem thus arbitrarily proposed; but it is interesting to know that there is no limit to the accuracy to which this ideal problem may be worked out by the methods which are actually used every day by engineers in their calculations and drawings.

The modern method of the "calculus of variations," brought into the perfect and beautiful analytical form in which we now have it by Lagrange, gives for this particular problem a theorem which would be very valuable to the draughtsman if he were required to produce an exceedingly accurate drawing of the required curve. The curvature of the curve at any point is convex towards the side on which the price per unit length of line is less, and is numerically equal to the rate per mile perpendicular to the line at which the Neperian logarithm of the price per unit length of the line varies. This statement would give the radius of curvature in fraction of a mile. If we wish to have it in yards we must take the rate per yard at which the Neperian logarithm of the price per unit length of the line varies. I commend the Neperian logarithm of price in pounds, shillings and pence, to our Honorary Secretary, to whom no doubt it will present a perfectly clear idea ; but less powerful men would prefer to reckon the price in pence, or in pounds and decimals of a pound. In every possible case of its subject the " calculus of variations " gives a theorem of curvature less simple in all other cases than in that very simple case of the railway line of minimum first cost, but always interpretable and intelligible according to the same principles.

Thus in Dido's problem we find by the calculus of variations that the curvature of the enclosing line varies in simple proportion to the value of the land at the places through which it passes ; and the curvature at any one place is determined by the condition that the whole length of the ox-hide just completes the enclosure.

The problem of Horatius Cocles combines the railway problem with that of Dido. In it the curvature of the boundary is the sum of two parts ; one, as in the railway, equal to the rate of variation perpendicular to the line, of the Neperian logarithm of the cost in time per yard of the furrow (instead of cost in money per yard of the railway); the other varying proportionally to the value of the land as in Dido's problem, but now divided by the cost per yard of the line which is constant in Dido's case. The first of these parts, added to the ratio of the money-value per square yard of the land to the money-cost per lineal yard of the boundary (a wall, suppose), is the curvature of the boundary when the problem is simply to make the most you can of a grant of as much land as you please to take provided you build a proper and sufficient stone wall round it at your own expense. This problem, unless wall-building is so costly that no part of the offered land will pay for the wall round it, has clearly a determinate finite solution if the offered land is an oasis surrounded by valueless desert.

It has also a determinate finite solution even though the land be nowhere valueless, if the wall is sufficiently more and more expensive at greater and greater distances from some place where there are quarries, or habitations for the builders.

The simplified case of this problem, in which all equal areas of the land are equally valuable, is identical with the old well-known Cambridge dynamical plane problem of finding the motion of a particle relatively to a line of reference revolving uniformly in a plane : to which belongs that considerable part of the " Lunar Theory " in which any possible motion of the moon is calculated on the supposition that the centre of gravity of the earth and moon moves uniformly in a circle round the sun, and that the motions of the earth and moon are exactly in this plane. The rule for curvature which I have given you expresses in words the essence of the calculation, and suggests a graphic method for finding solutions by which not uninteresting approximations[1]

---

[1] Kelvin, " On graphic solution of dynamical problems." *Phil. Mag.* 1892 (2nd half-year).

to the cusped and looped orbits of G. F. Hill[1] and Poincare can be obtained without disproportionately great labour.

In the dynamical problem, the angular velocity of the revolving line of reference is numerically equal to half the value of the land per square yard ; and the relative velocity of the moving particle is numerically equal to the cost of the wall per lineal yard in the land question.

But now as to the proper theorem of curvature for each case ; both Dido and Horatius Cocles no doubt felt it instinctively and were guided by it, though they could not put it into words, still less prove it by the "calculus of variations." It was useless knowledge to the bees, and, therefore, they did not know it ; because they had only to do with straight lines. But as you are not bees I advise you all, even though you have no interest in acquiring as much property as you can enclose by a wall of given length, to try Dido's problem

---

[1] Hill, *Researches in the Lunar Theory*, Part 3. National Academy of Sciences, 1887.

[2] *Méthodes Nouvelles de la Mécanique Céleste*, p. 109 (1892).

for yourselves, simplifying it, however, by doing away with the rugged coast line for part of your boundary, and completing the enclosure by the wall itself. Take forty inches of thin soft black thread with its ends knotted together and let it represent the wall ; lay it down on a large sheet of white paper and try to enclose the greatest area with it you can. You will feel that you must stretch it in a circle to do this, and then, perhaps, you will like to read Pappus (Liber V Theorema II. Propositio II.) to find mathematical demonstration that you have judged rightly for the case of all equal areas of the enclosed land equally valuable. Next try a case in which the land is of different value in different parts. Take a square foot of white paper and divide it into 144 square inches to represent square miles, your forty inches of endless thread representing a forty miles wall to enclose the area you are to acquire. Write on each square the value of that particular square mile of land, and place your endless thread upon the paper, stretched round a large number of smooth pins stuck through the paper into a drawing-board below it, so as to enclose as much value as you can, judging first roughly by eye and then correcting according to the sum of the values of complete squares and proportional values of parts of squares enclosed by it. In a very short time you will find with practical accuracy the proper shape of the wall to enclose the greatest value of the land that can be enclosed by forty miles of wall. When you have done this you will understand exactly the subject of the calculus of variations, and those of you who are mathematical students may be inclined to read Lagrange, Woodhouse, and other modern writers on the subject. The problem of Horatius Cocles, when not only the different values of the land in different places but also the different speed of the plough according to the nature of the ground through which the furrow is cut are taken into consideration, though more complex and difficult, is still quite practicable by the ordinary graphic method of trial and error. The analytical method of the calculus of variations, of which I have told you the result, gives simply the proper curvature for the furrow in any particular direction through any particular place. It gives this and it cannot give anything but this, for any plane isoperimetrical problem whatever, or for any isoperimetrical problem on a given curved surface of any kind.

Beautiful, simple, and clear as isoperimetrics is in geometry, its greatest interest, to my mind, is in its dynamical applications. The great theorem of least action, somewhat mystically and vaguely propounded by Maupertuis, was magnificently developed by Lagrange and Hamilton, and by them demonstrated to be not only true throughout the whole material world, but also a sufficient foundation for the whole of dynamical science.

It would require nearly another hour if I were to explain to you fully this grand generalisation for any number of bodies moving freely, such as the planets and satellites of the solar system, or any number of bodies connected by cords, links, or mutual pressures between hard surfaces, as in a spinning-wheel, or lathe and treadle, or a steam-engine or a crane, or a machine of any kind; but even if it were convenient to you to remain here an

hour longer, I fear that two hours of pure mathematics and dynamics might be too fatiguing. I
must, therefore, perforce limit myself to the two-
dimensional, but otherwise wholly comprehensive,
problems of Dido and Horatius Cocles. Going
back to the simpler included case of the railway of
minimum cost between two towns, the dynamical
analogue is this :—For price per unit length of the
line substitute the velocity of a point moving in a
plane under the influence of a given conservative
system of forces, that is to say, such a system that
when material particles not mutually influencing
one another are projected from one and the same
point in different directions, but with equal velocities, the subsequent velocity of each is calculable
from its position at any instant, and all have equal
velocities in travelling through the same place
whatever may be their directions. The theorem
of curvature, of which I told you in connection
with the railway engineering problem, is now
simply the well-known elementary law of relation
between curvature and centrifugal force of the
motion of a particle.

The motion of a particle in a plane is, as Liouville has proved, a case to which every possible
problem of dynamics involving just two freedoms
to move can be reduced. But to bring you to see
clearly its relation to isoperimetrics, I must tell
you of another admirable theorem of Liouville's,
reducing to a still simpler case the most general
dynamics of two-freedoms motion. Though not
all mathematical experts, I am sure you can all
perfectly understand the simplicity of the problem
of drawing the shortest line on any given convex
surface, such as the surface of this block of wood
(shaped to illustrate Newton's dynamical theory
of the elliptic motion of a planet round the sun)
which you see on the table before you. I solve
the problem practically by stretching a thin cord
between the two points, and pressing it a little this
way or that way with my fingers till I see and feel
that it lies along the shortest distance between
them. And now, when I tell you that Liouville
has reduced to this splendidly simple problem of
drawing a shortest line (geodetic line it is called)
on any given curved surface every conceivable

problem of dynamics involving only two freedoms to move, I am sure you will understand sufficiently to admire the great beauty of this theorem.

The doctrine of isoperimetrical problems in its relation to dynamics is very valuable in helping to theoretical investigation of an exceedingly important subject for astronomy and physics—the stability of motion, regarding which, however, I can only this evening venture to show you some experimental illustrations.

The lecture was concluded with experiments illustrating—

1. Rigid bodies (teetotums, boys' tops, ovals, oblates, &c.) placed on a horizontal plane, and caused to spin round on a vertical axis, and found to be thus rendered stable or unstable according as the equilibrium without spinning is unstable or stable.

2. The stability or instability of a simple pendulum whose point of support is caused to vibrate up and down in a vertical line, investigated mathematically by Lord Rayleigh.

3. The crispations of a liquid supported on a

vibrating plate, investigated experimentally by Faraday; and the instability of a liquid in a glass jar, vibrating up and down in a vertical line, demonstrated mathematically by Lord Rayleigh.

4. The instability of water in a prolate hollow vessel, and its stability in an oblate hollow vessel, each caused to rotate rapidly round its axis of figure,[1] which were announced to Section A of the British Association at its Glasgow meeting in 1876 as results of an investigation not then published, and which has not been published up to the present time.

[1] *Nature*, 1877, vol. 15, p. 297, "On the Precessional Motion of a Liquid."

INDEX.

R R

END OF VOL. II.

RICHARD CLAY AND SONS, LIMITED, LONDON AND BUNGAY.

Printed in the United States
By Bookmasters